IET CONTROL, ROBOTICS & SENSORS SERIES 99

Practical Robotics and Mechatronics

Also available and coming soon:

Volume 68	**Flexible Robot Manipulators: Modelling, simulation and control** M.O. Tokhi and A.K.M. Azad (Editors)
Volume 69	**Advances in Unmanned Marine Vehicles** G.N. Roberts and R. Sutton (Editors)
Volume 77	**Further Advances in Unmanned Marine Vehicles** G.N. Roberts and R. Sutton (Editors)
Volume 86	**Modeling and Control of Flexible Robot Manipulators** M.O. Tokhi and A.K.M. Azad.
Volume 96	**Cyber-Physical System Design with Sensor Networking Technologies** S. Zeadally and N. Jabeur (Editors)
Volume 97	**Sensory Systems for Robotic Applications** R. Dahiya (Editor)
Volume 104	**Control of Mechatronic Systems** L. Guvenc, A. Guvenc, B. Demirel and M.T. Emirler
Volume 105	**Mechatronic Hands: Prosthetic and Robotic Design** P.H. Chappell

Practical Robotics and Mechatronics

Marine, Space and Medical Applications

Ikuo Yamamoto

The Institution of Engineering and Technology

Published by The Institution of Engineering and Technology, London, United Kingdom

The Institution of Engineering and Technology is registered as a Charity in England & Wales (no. 211014) and Scotland (no. SC038698).

First published 2016

The Institution of Engineering and Technology
Michael Faraday House
Six Hills Way, Stevenage
Herts, SG1 2AY, United Kingdom

www.theiet.org

British Library Cataloguing in Publication Data
A catalogue record for this product is available from the British Library

ISBN 978-1-84919-968-1 (hardback)
ISBN 978-1-84919-969-8 (PDF)

Typeset in India by MPS Limited

Contents

Preface

The world is experiencing the beginning of a revolution in robotics and mechatronics. A key part of this revolution is integration with IoT (Internet of Things), this networking of robotics and mechatronics will represent significant market opportunities.

The author engages in developing technologies using IoT-based intelligent robotics using several effective applications as examples. The successful development of practical robotics and mechatronics must begin from consideration of IoT and/or the M2M (Machine To Machine) interface.

The author has developed many world-class practical examples of robotics and mechatronics during his 30 years of experience as an engineer and a professor. The author's experience comprises working with Mitsubishi Heavy Industries Ltd. (20 years), JAMSTEC (3 years), and as a professor at Kyushu, Kitakyushu, and Nagasaki University (10 years). Most of the robotics and mechatronics developed continue to be used safely. Such an experience of development inherently creates new ideas for robotic applications. Based on the author's experience, important ways of thinking and technologies essential to produce successful robotics and mechatronics are summarized in this book.

The author hopes that this book can provide fruitful hints for developing new robotics and mechatronics that will lead to new and successful business opportunities.

Ikuo Yamamoto, Dr. Eng.,
Nagasaki, Japan

Chapter 1
Introduction

There is a high demand for developing robotics and mechatronics around the world. The market for robotics and mechatronics continues to expand, and is expected to be an increasing business field. The market needs practical robotics and mechatronics, which meet societal demands.

How can we develop a new practical robotics and mechatronics?

The author has developed practical robotics and mechatronics during his 20 years or so at Mitsubishi Heavy Industries, Ltd., and has invented many products, such as a 10000-m operating depth capable unmanned remotely operated vehicle (ROV), called Kaiko; a manned underwater vehicle, the Shinkai 6500; a hydrofoil catamaran called Rainbow; an air cushion surface effect ship, the Techno Super Liner; the intelligent ship, the Super Joy; a dynamic positioning system (DPS) of offshore platform, like the bridge construction vessels UMASHIMA and CHIKYU; a riser entry system, robotic fish; the B787 main wing, a regional jet. The author was the project leader of the autonomous underwater vehicle (AUV) Urashima, which established a world record for autonomous continuous cruising in 2005. In addition, as a university professor for over 10 years, the author has developed intelligent robotics such as robotic fish, a multirotor aviation system, medical robotics, and space robotics. Particularly – true story! – one of the author's robotic fish swam in space.

This book can provide hints for solutions to developing new robotics and mechatronics which are intelligent and practical.

Robotics is defined as 'an intelligent machine system which consists of sensors, actuators, and a controller'. If robotics is compared to humans, the relationship of sensors, actuators, and a control device can be summarized as shown in Figure 1.1.

Sensors are the devices that feel or sense the outer and inner states of the robot, similar to the eyes, ears, nose, and pressure/temperature points of the skin in a human body. Actuators are the devices that actuate parts of the body, just like the hands, feet, and mouth. A controller is the device that makes decisions according to the sensor information and other data, and gives commands to the actuators to move (or stop), as in the brain of a human.

Mechatronics is defined as 'mechanics with an electrical circuit'. Mechatronics is originally a Japanese-made English word, which is a mixture of mecha(nics) and (elec)tronics. Robotics is composed of mechatronics, as shown in Figure 1.2.

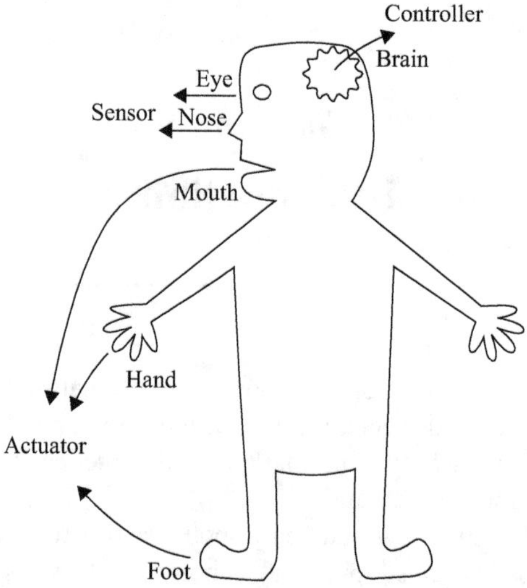

Figure 1.1 Robotic functionality compared to humans

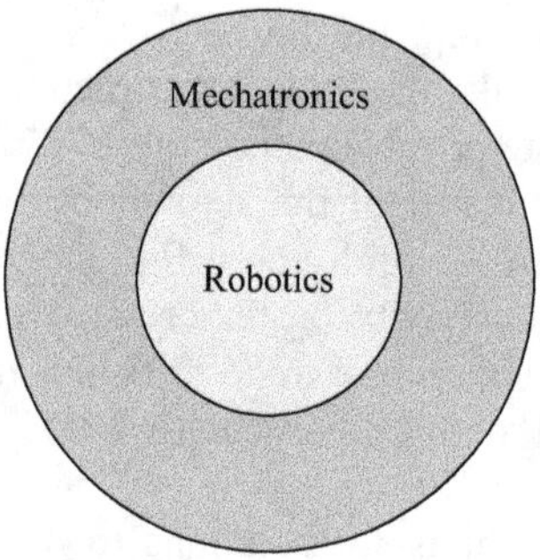

Figure 1.2 Relationship between robotics and mechatronics

Robotics and mechatronics consist of many fundamental technologies, such as mechanics, electronics, electrical engineering, and information technology. Fundamental technologies sometimes come from economics, medicine, agriculture, fishery, literature, and other scientific subjects. The author refers to such technologies as 'universal knowledge', where 'universal' provides the root word of 'university'. That is, the knowledge important for robotics and mechatronics are subjects which students learn during their undergraduate terms at university.

The basic procedure to create new robotics and mechatronics can be summarized as shown in Figure 1.3.

Fundamental technologies and their integrated systems produce seeds for robotic and mechatronic inventions. Customers outline their needs, and the result is the development of new robotics and mechatronics.

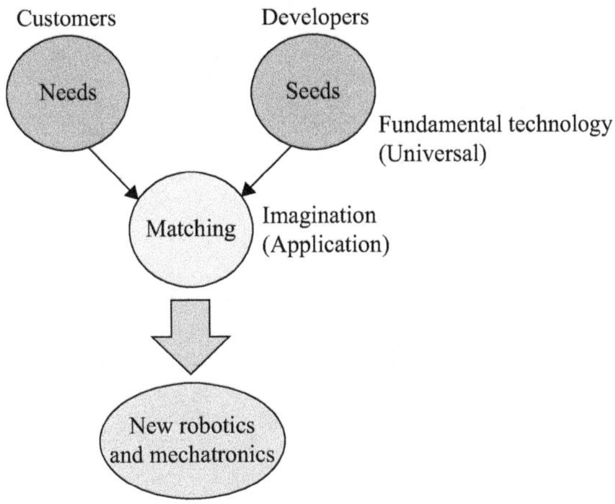

Figure 1.3 Basic procedure to create new robotics and mechatronics

It is important to develop new robotic and mechatronic mechanisms to meet customers' requests, and to be creative and highly functional. In addition, the Japanese methods of manufacturing are well suited to the development of new robotics and mechatronics.

There is a Japanese word, *osekkai*, which means 'interference and meddlesomeness from goodwill'. Development of robotics and mechatronics usually involves group work, and it is important that each engineer is not only professional in one element of technology but also mindful of other fields of technology, and entirely covers all fields of robotic and mechatronic development in the group.

Chapter 2

Importance of robotics and mechatronics in society

Classification of robotics and mechatronics, and their needs in society are outlined in this chapter.

Robotics and mechatronics penetrate into every part of society. The definition of robotics is as follows: 'Robotics is a system of intelligent machinery which consists of sensors, actuators, and a controller.' The definition of mechatronics is 'mechanics including an electrical circuit'.

You can find various robotics and mechatronics in society. If you look around the room, you can find several examples of robotics and mechatronics. For example, an air conditioner is robotic. It is an intelligent machine which has a sensor (thermometer), actuator (air emission machine), and controller (microcomputer for adjustment of temperature). Similarly, a vacuum cleaner is robotic, or at least mechatronic. Some recent vacuum cleaners can avoid obstacles automatically by sensing and conduct a cleaning mission by itself.

If you go to an airport, you sometimes find a human-like robotic guide. The boarding gate, which does not look like a human, is very robotic. It can sense barcode information of boarding slips and guide passengers to the airplane automatically.

There are various types of robotics and mechatronics in society, and robotics and mechatronics help people with their daily lives. Generally, robotics can be classified into the following categories:

- 'Service robots', which focus on providing convenience. Robotic navigation for local directions, car navigation guidance robots in town, and in-vehicle navigation robots belong to this category.
- 'Amusement robots', which entertain people. Humanoid robots, dog robots, and fish robots, which were created earlier in Japan, belong to this category.
- 'Industrial robots', which work in factories carrying out hazardous manufacturing operations and monotonous work in place of people. A welding robot on an automobile manufacturing line is representative of this category. Industrial robots that focus on specific tasks, such as a welding robot, provide only a human arm- and hand-like function.

Chapter 3

How to create practical robotics and mechatronics

The experiential methods and basic ideas to create practical robotics and mechatronics are presented in this chapter.

Robotics and mechatronics are based on several technologies. The most important technologies are mechanics, electronics, electrical engineering, and information technology. A prominent person sometimes creates and develops robotics and mechatronics alone; however, it is rare for a single person to produce them to a marketable level. Usually, a team is organized by engineers from mechanics, electronics, software, manufacturing, and business persons who direct the product development to the market. The author has experienced, that the optimal number of people to create new robotics and mechatronics is five.

The individual and team experience in developing robotics and mechatronics becomes 'treasure', which contains both successes and failures. The development of one field of robotics and mechatronics leads to the development of other fields; this means they produce cluster technologies that are an accumulation of technological development and result in increased business in various markets. Development of industrial products is divided into two groups, which are 'sustainable development' and 'evolutional development'. Sustainable development is intended to improve and modify established industrial products, whereas evolutional development creates new technologies to surpass conventional products.

Time histories of evolution of both developments can be summarized as in Figure 3.1. Sustainable development is based on significant knowledge and experience. It sometimes encounters failures during development, but they can be solved quickly with experience. Evolutional development has potential of high evaluation, but it involves risk of failure due to a lack of experience. If substantial failure occurs during development, the evaluation becomes zero and developers are compelled to suspend or stop the development. The important thing to do in this case is to maintain 'positive motivation' and carefully analyse the cause of failure.

The author invented a life-like sea bream robotic fish in 1995 (the first of its kind in the world), the first robotic coelacanth in 1997, and a robotic whale in 1998; however, we suffered a failure when the new robotic coelacanth was demonstrated to the public in 2000. We could not know the reason why it did not swim at the time of presentation, and the evaluation went down to zero. Afterwards, through failure analysis, we found that the reason was a heat problem in the battery

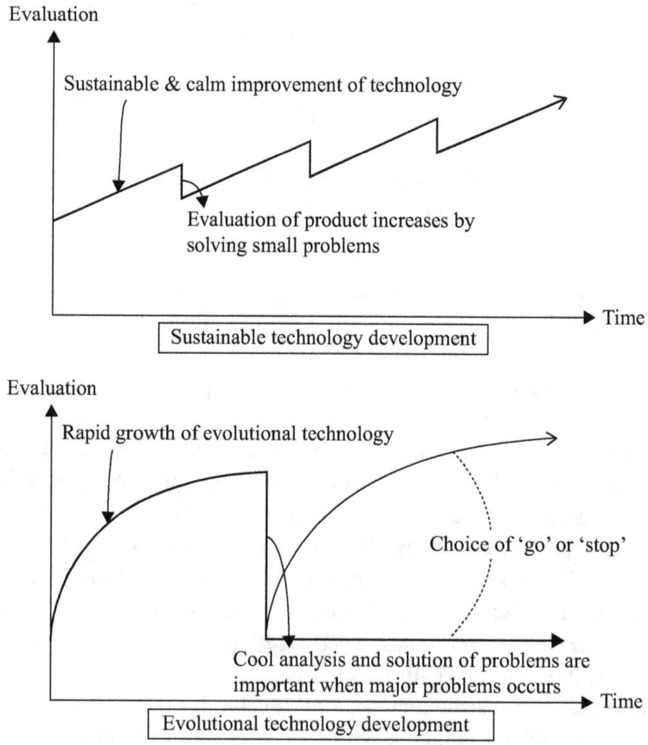

Figure 3.1 Sustainable technology and evolutional technology

caused by the TV cameras' strong light. The skin of the robotic fish around the battery was very thin and influenced by the heat of light. On the basis of this failure we have produced more robust robotic fish resulting in receiving a higher level of evaluation.

The author developed over 16 kinds of robotic fish and the fundamental mechanism went through four evolutions of change. The author encounters failures when a new mechanism is used; however, the problems can be overcome by positive motivation and a cool theoretical mind.

The newly developed robotics and mechatronics are called a 'prototype' model. Production of the actual products starts after confirmation of specification through practical tests of the prototype model. Most of the products which the author developed have been successful. The author has the experience and principles shown in Figure 3.2 according to the various successes and occasional failures.

That is, development of a new model starts over the wall of possibility (which is 3% of completeness), development of products is fully in progress over the wall of realization (which is 70% of completeness), and complete over the wall of reliance (which is 99% of completeness). The products are on the market after clearing all three walls. A feasibility study (FS) is conducted from the wall of possibility to that

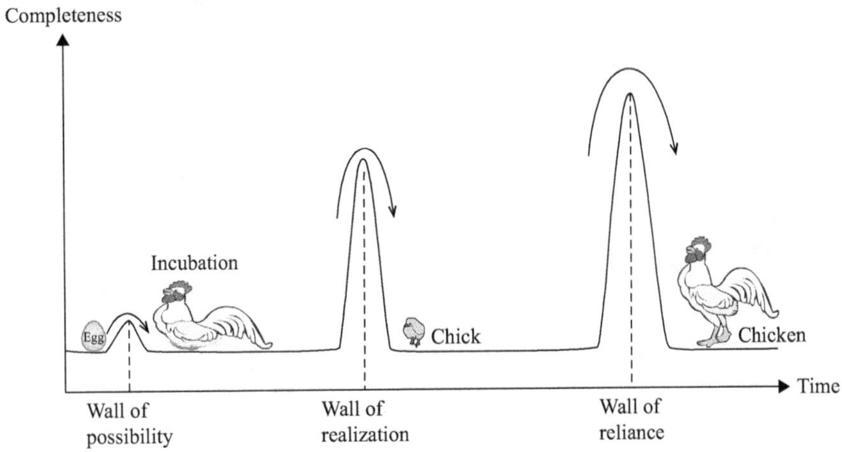

Figure 3.2 Three walls for product development

of realization. If the product can't be realized, the technical experience gained can be used to face the next challenge. If a product's probability of success exceeds 70% (the wall of realization) it should proceed to the manufacturing stage. Nothing in the world is produced without facing some kind of challenge.

Every new product faces the reliance wall at approximately 99% of completeness.

The product should satisfy the performance specification. The author has had many experiences of going over the walls of possibility realization and reliance. For example, when we achieved the long continuous autonomous cruising over 300 km by the AUV Urashima, the developers faced and overcame hardware and software problems in severe environmental conditions and completed all missions. The author experienced significant mental fatigue; however, the last wall of reliance must be overcome in order to realize the completion of robotics and mechatronics.

Chapter 4

Basic system design procedure for robotics and mechatronics

Successful methods of designing basic systems for robotics and mechatronics are described in this chapter.

Robotics and mechatronics are a 'system'. The definition of system is a 'network', which consists of 'multiple elements, and has a net performance that exceeds that of each element'. Robotics is a system which consists of sensors, actuators, and a controller, according to the definition of the Japanese Ministry of Economics. Engineers in Japanese industries in the twentieth century referred to mechatronics as a system which consists of mechanics and electronics.

A generalization of the construction method of robotics and mechatronics is summarized in Figure 4.1.

The process starts by planning a product, then basic design and detailed design are carried out, and this proceeds to manufacturing, operational testing, and maintenance.

The basis for development of the system is 'simulation in advance by computer and/or manually, based on mathematical models and/or dynamics and model tests before manufacturing products'. This is very important for avoiding high risks and costly failures. Logistic approach by dynamics and solving problems from small

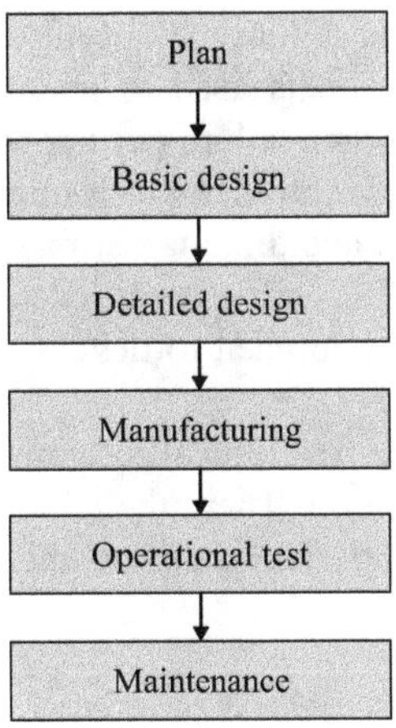

Figure 4.1 System construction flow of robotics and mechatronics

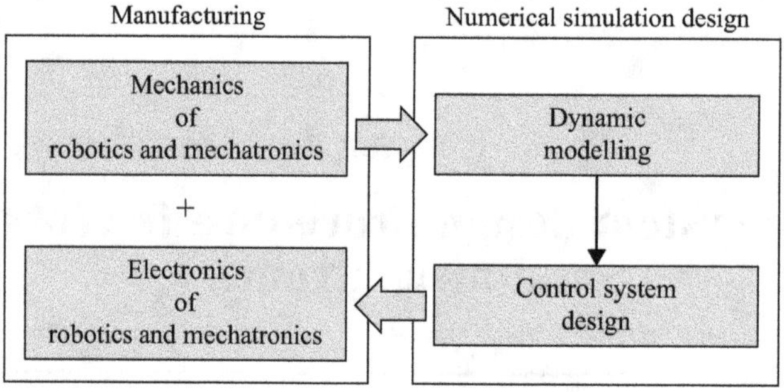

Figure 4.2 Relation between manufacturing and numerical simulation design

failures by dynamic model tests at early stages are key for successful development of products. If a fatal failure is found at the final stage of development and after manufacturing, the cost for rectifying the problem may be very high and it results in significant management pressure from the business. Numerical simulation and model experiments are conducted during basic design at the planning stages to reduce the risk of failure. The cost at the initial stages by simulation is a small part of the total product's development cost.

Recently, the use of computer-aided design (CAD) has become popular during the design stage. It is a common way to make a three-dimensional (3-D) draft with numerical static/dynamic simulations. In addition, there is an increasing use of computer-aided manufacturing (CAM) to automate the manufacture of products. This method contributes to a reduction of production costs and an increase in the stabilization of a product's quality, particularly in case of mass production.

The author has experience in meeting product specifications by solving technical problems after manufacture. The reason for trouble is sometimes simple, such as a software bug or the quality of a machine part. However, every development team has apprehension that total failure may occur due to one small failure during the development. Fail-safe design is used to maintain operation by substituting other parts during the mission when failure occurs, which is why it is important for the development team to consider problems that might occur during the operation. In addition, foolproof design to avoid human error should also be considered.

Robotics and mechatronics needs both mechanical and electrical system manufacturing. The successful approach to development for robotics and mechatronics is summarized in Figure 4.2.

Dynamic modelling of the mechanical system and control system design is conducted by numerical simulation. Then, the controller included in the designed control software is equipped with the electronic and electrical system.

Chapter 5

Dynamics and control of robotics and mechatronics

The dynamics and control design necessary in the development of robotics and mechatronics are reviewed in this chapter.

Why are new theories using advanced control not applicable for practical robotics and mechatronics? Almost all control systems of robotics and mechatronics are designed by classical control theory. That is, sequential control by 'IF-THEN-ELSE' logic and feedback control using proportional-integral-derivative (PID) control can be easily used by engineers to meet control specifications. However, simpler control logic may be preferable, as follows:

If you develop a collision avoidance function for a robotic fish, you place the sonar sensor at the nose of the robotic fish and design the following logic.

(IF)The sonar sensor detects an obstacle in the forward direction, (Then) turn by rotating several degrees of the fin. (Else) if it finds nothing, go forward as is. It is a practical design method for obstacle avoidance by sequential control.

If you design depth control in a robotic fish, the depth sensor (pressure sensor) is placed inside the robotic fish and the logic shown in Figure 5.1 is designed.

If the robotic fish swims at a depth of 10 m, you set up the reference command at 10 m. The depth sensor of the robotic fish detects the present swimming depth and provides feedback to the reference command. Deviation between the reference command and the present status is calculated, and the actuator controls the valve of the buoyance device by adjusting the inlet or outlet of air to achieve zero deviation using a PID controller. PID, PI, PD, and P controllers are used according to the characteristics of the mechanism to be controlled.

PID control is a feedback system. A feedforward system is sometimes used in conjunction with a feedback system, when the time response for setting up a reference

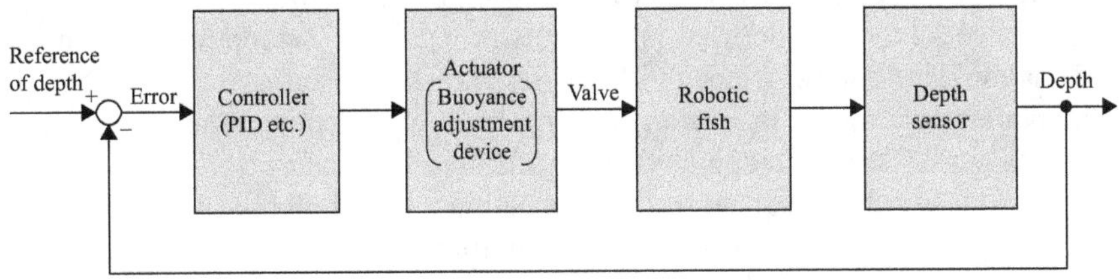

Figure 5.1 Feedback control system of depth for a robotic fish

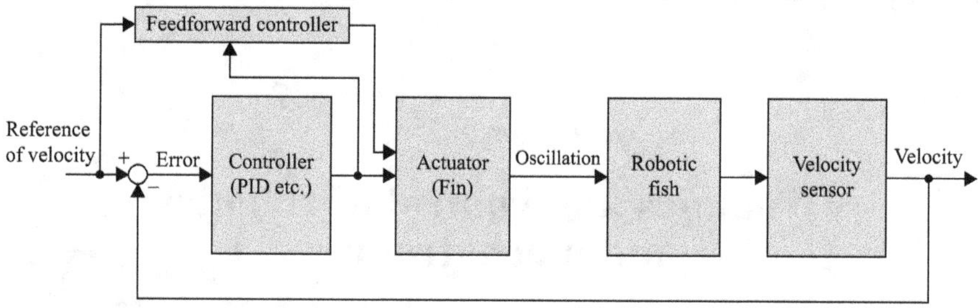

Figure 5.2 Feedback control system of velocity control for a robotic fish

command needs to be improved. Figure 5.2 shows a sample of feedforward control to control the velocity of a robotic fish. The feedforward system learns how to give control signals to the actuator in order to satisfy the reference command from the feedback system. Finally, it is possible to give commands to the actuator promptly without a feedback system, significantly improving the time response of the system. If the robotic fish swims at 5.0 km/h (quickly), a feedforward system learns the relationship between velocity and the oscillating fin parameters, such as frequency, amplitude, and phase difference from the feedback system, and promptly gives commands to the oscillating fin to swim at 5.0 km/h.

Researchers and engineers should begin the development of robotics and mechatronics using the above-mentioned simple control algorithm. They should gain experience in working with robotics and mechatronics as soon as possible, particularly with regard to seeing a project through to completion. Control engineers in particular should gain experience in the practical behaviour of robotics and mechatronics. Sequential control and PID control should be the first applications of robotics and mechatronics.

The controlled mechanism can be classified as a single-input, single output (SISO) or multiple-input, multiple-output (MIMO) system, as shown in Figure 5.3.

A SISO system is a fundamental model of motion and can be controlled by a PID control system. In a MIMO system, according to the author's experience, PID control can be applicable for up to two inputs and two outputs. If the system has more than two inputs and two outputs, regulator theory is best in order to determine stable controller gains. The engineers should make the dynamics of the controlled mechanism and modelling of state equations and output equation using linearization of the dynamics before considering the use of a regulator.

If you make a controller that is robust with regard to disturbances and model the deviation, H∞ control which produces robustness for plant, is recommended. In addition, the linear matrix inequality (LMI) method of modelling assures modelling which has nonlinearity and applications to adaptive H∞ controls.

If you face problems of control for a small number of inputs and outputs, and the controlled mechanism's behaviour has significant nonlinearity (which refers to unpredictable behaviour), adaptive control using sliding mode control, self-tuning

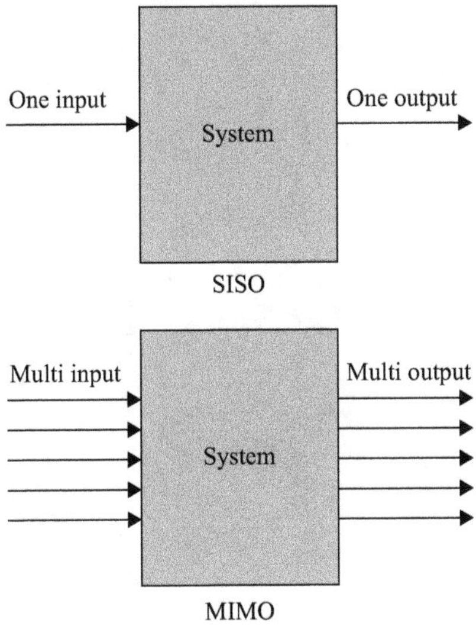

Figure 5.3 SISO and MIMO systems

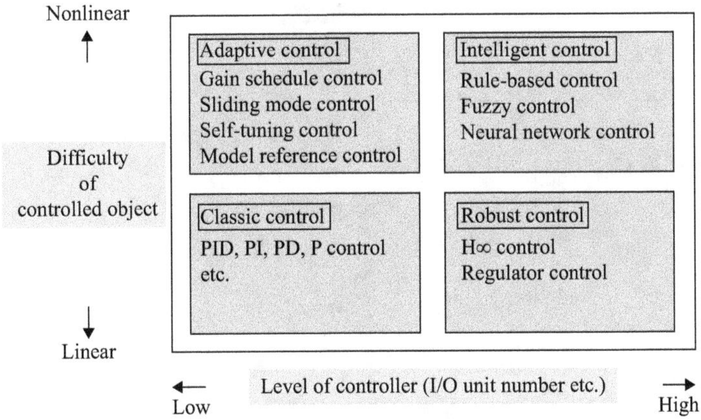

Figure 5.4 Principle of control system design method

adaptive control, model reference type adaptive control, and gain scheduling control approaches are applicable.

If you have problems with a MIMO system and high nonlinearity of the controlled mechanism, fuzzy control, neural network control, and rule-based control are recommended for such applications.

An application map of the control system can be summarized as shown in Figure 5.4.

Chapter 6

Advances in marine robotics and mechatronics

6.1 Autonomous underwater vehicle

6.1.1 Overview

The past, present, and future AUV technologies developed by the author, who belonged to the Japan Agency for Marine-Earth Science and Technology (JAMSTEC) as the leader of the autonomous underwater vehicle group, are described in this chapter. The first to be presented is the deep and long-distance cruising AUV Urashima, powered by a fuel cell, with the following specifications: 10-m length; 2.5-m width; 2.4-m height; 10-tons weight; 300-km cruising distance; and 3500-m depth diving at 3-kt speed autonomously in the ocean. It is used for sea bottom surveys in earthquake areas and for research of global warming phenomena. The second AUV technology described is the next-generation AUV designed by the author for higher manoeuvrability and longer distance cruising than Urashima. Third, this chapter details biomanoeuvring AUVs, which are fish-like swimming robots used for various purposes in scientific research. Finally, a marine system network based on AUVs is described.

6.1.2 Long-distance cruising AUV

6.1.2.1 Design concept

The deep and long-distance cruising AUV Urashima, which can be used for oceanographic surveys such as sea bottom surveys in earthquake areas and the research of global warming phenomena, was developed by JAMSTEC. The principal specifications of the AUV Urashima are listed in Table 6.1. It is designed to have a maximum cruising range of 300 km and a maximum working depth of 3500 m, powered by a fuel cell.

Urashima is equipped with vertical thrusters, a horizontal thruster, horizontal rudders, and vertical rudders for motion control; a buoyancy adjustment system, an oil bladder, a trim adjustment system, and ballast for buoyancy adjustment; and a high-performance lithium-ion battery as a secondary energy source that can be used alone for cruising to some extent. In addition, a digital camera and side-scan sonar are installed for precision seafloor exploration. Figure 6.1 shows a picture of Urashima developed by the author when the world record for cruising was established. Figure 6.2 shows the general arrangement of Urashima.

Table 6.1 Specifications of AUV Urashima

Length overall	10 m	**Power source:** Fuel cell (PEFC) and lithium-ion rechargeable battery
Width overall	2.5 m	
Height overall	2.4 m	
Width (shell)	1.3 m	**Navigation instruments:** Inertial navigation system, acoustic doppler, current profiler, acoustic homing sonar, obstacle avoidance sonar
Height (shell)	1.5 m	
Weight in the air	10 ton	
Maximum working depth	3500 m	
Maximum cruising range	300 km	**Observation and exploration instruments:** Colour TV camera, CTDO sensor, side-scan sonar, automatic water sampler, digital camera
Cruising speed	3 kt	
Maximum speed	4 kt	

Figure 6.1 Deep-sea cruising AUV Urashima

As a main power source, the fuel cell system, which has high energy density, was adapted for the AUV Urashima. The fuel cell for the AUV generates electricity by electrochemical catalysis of hydrogen and oxygen. The hydrogen is stored safely within a metal hydride hydrogen-absorbing alloy.

In June 2004, Urashima successfully achieved an autonomous mission of 220 km in Suruga Bay, Shizuoka, Japan. It made four roundtrips between transponders, spending 44 h submerged. In the February 2005 sea trial, Urashima achieved

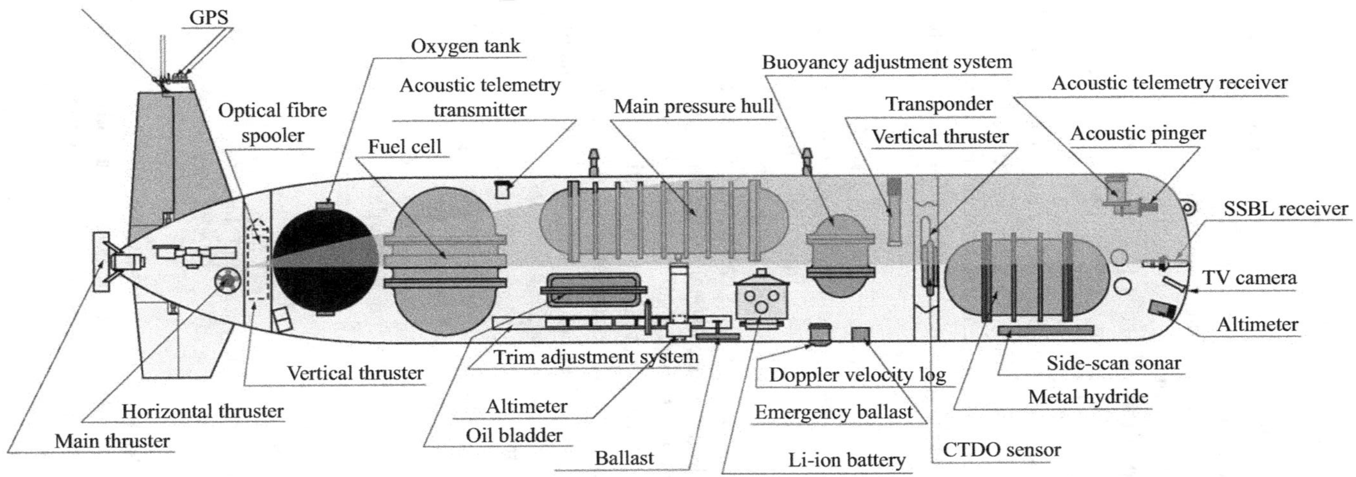

Figure 6.2 General arrangement of AUV Urashima

an autonomous long-range cruise to the world record of 317 km for a fuel-cell-powered AUV.

6.1.2.2 Battery

Lead acid and silver zinc batteries have traditionally been used as power sources for underwater vehicles. Increased electric capacity is needed for extended cruising and requires an increase in battery volume and weight, resulting in a greatly enlarged vehicle. As a long-distance cruising AUV requires a power source capable of operating for a long period of time, it needs an air-independent propulsion (AIP) power source characterized by high energy density and high energy efficiency. Thus, a power source with high energy density and high energy efficiency is an essential ingredient to reduce the size of the vehicle. Although prototype fuel cells, Stirling engines, closed-cycle diesel engines (CCDE), and closed steam turbines have been developed and manufactured in recent years as underwater power sources, a fuel cell was selected as Urashima's most suitable power source for the following reasons.

Fuel cell

Fuel cells have recently received attention as an effective means of generating power and solving energy and environmental problems. Fuel cells could take over internal combustion engines in the near future, eliminating roadside exhaust gasses. In the future, more efficient fuel cell unit generators will supply electric power and hot water for residential utilities, and easily rechargeable fuel cells may replace batteries in cellular phones and personal computers. This eliminates the need for charging and recharging batteries. Fuel cell technology is relatively old, having been invented in 1839 by British physicist Sir William Grobe, the father of the fuel cell. The fuel cell procedure reverses the process of water electrolysis. In a fuel cell, hydrogen and oxygen react electrochemically to produce electric power. Conventional internal and external combustion engines first change thermal energy to mechanical energy, and then into electric energy. Fuel cells, on the other hand, change chemical energy directly into electric energy, and therefore have higher energy conversion efficiency. Fuel cells are especially effective in reducing CO_2 emissions. If hydrogen is used as fuel and oxygen is used as an oxidizer, the reactive product emitted from the fuel cell will be only water. Because hydrogen can be produced in various ways, fuel cells are a highly effective power-generation method that can solve energy problems. Fuel cells are the most suitable AUV power source for the following reasons:

- High power generating efficiency and less fuel and oxidizer consumption compared with other power sources. This can minimize the volume and weight of the vehicle.
- Silence is a necessity for AUVs because they carry acoustic equipment such as an acoustic communication system, obstacle avoidance sonar, and side-scan sonar. Fuel cell stacks have no reciprocities or rotational moving parts and are thus superior in terms of sound level emission.
- Fuel cells emit only water as a result of the hydrogen/oxygen reaction. This allows for easy storage and discharge.

Table 6.2 Types of fuel cells

Type	**Phosphoric acid Fuel cell (PAFC)**	**Molten carbonate fuel cell (MCFC)**	**Solid oxide fuel cell (SOFC)**	**Polymer electrolyte fuel cell (PEFC)**
Electrolyte	Phosphoric acid	Carbonate	Solid oxide	Protein exchange membrane
Operating temperature	150–200 °C	600–650 °C	800–1000 °C	up to 80 °C
Fuel	Natural gas, LPG, etc.	Natural gas, coal gasification gas, etc.	Natural gas, coal gasification gas, etc.	Hydrogen, natural gas, methanol, etc.
Exhaust heat utilization	Hot water, steam	Gas turbine, steam turbine	Gas turbine, steam turbine	Hot water
Features	Low temperature operation	Extensive use of fuel/internally reformable	Extensive use of fuel/internally reformable	High power density, low temperature operation

Table 6.3 Fuel cell system specifications

Items	**Specifications**
System	Closed-cycle system
Fuel cell type	PEFC (Polymer electrolyte fuel cell)
Output	DC 4 kW
Voltage	120 V
Number of stacks	2
Fuel	Pure hydrogen (Metal hydride storage)
Oxidizer	Pure oxygen (High-pressure tank storage)

Polymer electrolyte fuel cell

Depending on the type of electrolyte, fuel cells are classified into phosphoric acid, molten carbonate, solid oxide, or polymer electrolyte fuel cells (PEFCs). In the types shown in Table 6.2, PEFC fuel cells are smaller and have a lower operating temperature compared to other types of fuel cells. They are better suited for distributed power supplies, household power supplies, and automotive power. The development of PEFCs has advanced considerably in recent years. The PEFC was adopted as the Urashima's power source for the same reasons. As the operating temperature is close to room temperature, it allows for easy starting and stopping. It is also possible to use thermal insulation and build a smaller and lighter system configuration.

Urashima fuel cell system

Table 6.3 shows the specifications of the Urashima fuel cell system, and Figure 6.3 shows the Urashima fuel cell system. The Urashima fuel cell system has two

Figure 6.3 The fuel cell system and the author

stacks connected in series and have an output of 4 kW at 120 V. Figure 6.4 shows a diagram of the Urashima fuel cell system. The Urashima fuel cell system is a completely closed-cycle fuel cell, unlike the fuel cells used on land. The fuel cell system is installed in a pressure vessel. Hydrogen is supplied from a metal hydride hydrogen storage system installed in another pressure vessel. Oxygen is supplied from a high-pressure oxygen tank. Unreacted gas that has passed through the fuel cell stacks is re-circulated and the generated water is stored in the reaction water tank installed under the fuel cell pressure vessel. Temperature and humidity controls are very important in PEFCs. Operating temperatures are controlled by changing the circulating water flow rate to the cooler. Operating humidity is controlled by the humidifier. Temperature and humidity controls are maintained by thermal insulation and heating of the gas lines. A 30-h lithium-ion rechargeable battery is also installed in order to assist the fuel cell system if and when necessary. The lithium-ion rechargeable battery can supply sufficient current when required electricity demand exceeds the fuel cell's maximum capacity due to a temporary overload.

Hydrogen storage

One of the main considerations with fuel cells is hydrogen storage. Table 6.4 shows various proposed hydrogen storage methods. Taking into account the feasibility and

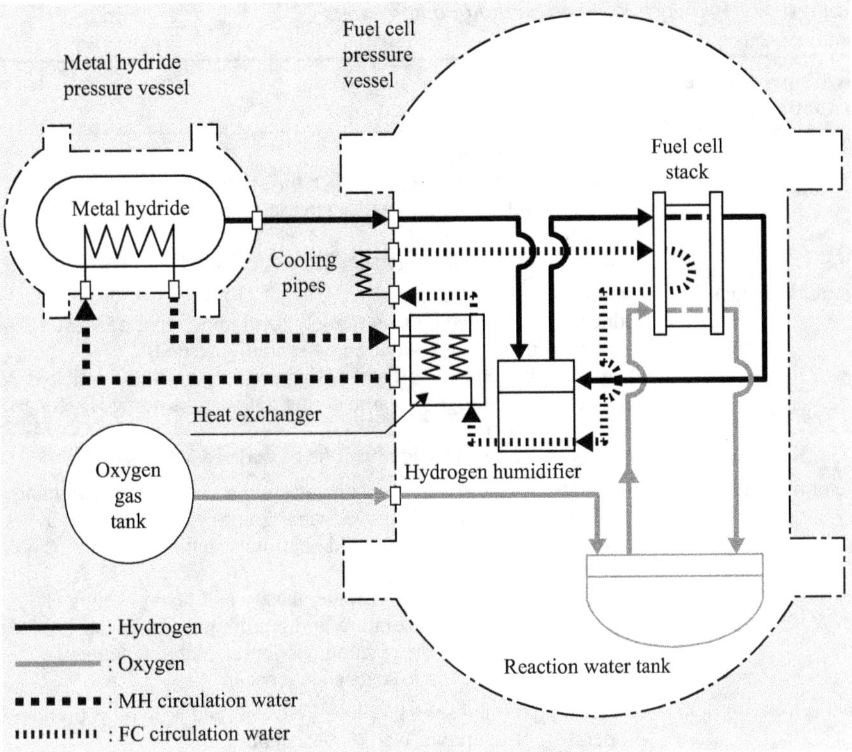

Figure 6.4 Fuel cell diagrams

adoptability of hydrogen storage in Urashima, a newly developed 30-MPa high pressure hydrogen tank was initially planned to be used. However, after researching other methods, metal hydride hydrogen storage has been adopted as a safer method.

Urashima metal hydride hydrogen storage system

Metal hydride desorbs hydrogen with an endothermic reaction. This reaction decreases the hydride temperature and equilibrium hydrogen pressure during desorption. As a result, the reaction restricts any large volume of hydrogen leakage if a leak should occur. As a safer method of hydrogen storage, metal hydride storage was adopted for Urashima. After several types of metal hydrides had been examined at the preliminary stage, an AB5-type rare-earth-based alloy was finally chosen for the Urashima metal hydride storage system on board. It exhibits a good absorption efficiency and is easy to use because its absorption temperature is 10–20 °C (room temperature) and the desorption temperature is 40–60 °C (fuel cell operating temperature) (Figure 6.5). Hydrogen desorption is accelerated using fuel cell waste heat. The metal hydride hydrogen storage system consists of a metal hydride storage container, circulating water pump, sensors, and controller in the pressure vessel.

Table 6.4 Hydrogen storage method comparisons

Hydrogen storage method	Features
High-pressure gas	– gaseous state storage – gas can be compressed to reduce the storage space – vessel weight increases with increase of gas pressure – practical use on land – safety in time of leakage should be considered
Liquid hydrogen	– stored in a liquid state – due to storage method at extremely low temperature of −250 °C or lower (a insulating structure is essentially needed) – liquid hydrogen boils off and arising pressure must be ventilated, so that storage hydrogen volume during long period storage is reduced. – practical use on land – safety in case of leakage must be considered
Metal hydride	– high volumetric storage density and compact storage are available – hydrogen storage rate changes according to alloy type – hydrogen desorption is an endothermic reaction and heat should therefore be applied – presents low gravimetric storage density and heavy weight – decreases hydride temperature and equilibrium hydrogen pressure while desorption, so the reaction restricts any large volume of hydrogen leakage if a leakage should occur.
Fuel reforming	– reforming of methanol, gasoline, kerosene, and light oil is proposed – operating temperature is 300 °C or more – stored in liquid and thus easy to handle – produces byproducts including CO, CO_2
Chemical hydride	– at the laboratory research stage, not yet practically feasible – more research and development is necessary before practical applications can be realized – regeneration requires a significant amount of energy
Carbon nano-tube/ ullerene	– at the laboratory research stage, not yet practically feasible – more research and development is necessary before practical applications can be realized – at present production cost is extremely high

6.1.2.3 Control system

For designing the Agent-Net-based control system, a macro design and a micro design must be considered. The macro design means that each vessel is considered as one agent, and that each agent must operate autonomously to achieve the overall goal of the system. The micro design means that each control software component is considered as one agent, and that the agents must cooperate autonomously to achieve the aim of the vessel. In such systems a hierarchical control system is important to the software design.

The control system software for an AUV is based on two main design principles, that is, hierarchization and decentralization of the decision-making. In order to make the control system robust enough for long-distance autonomous navigation in

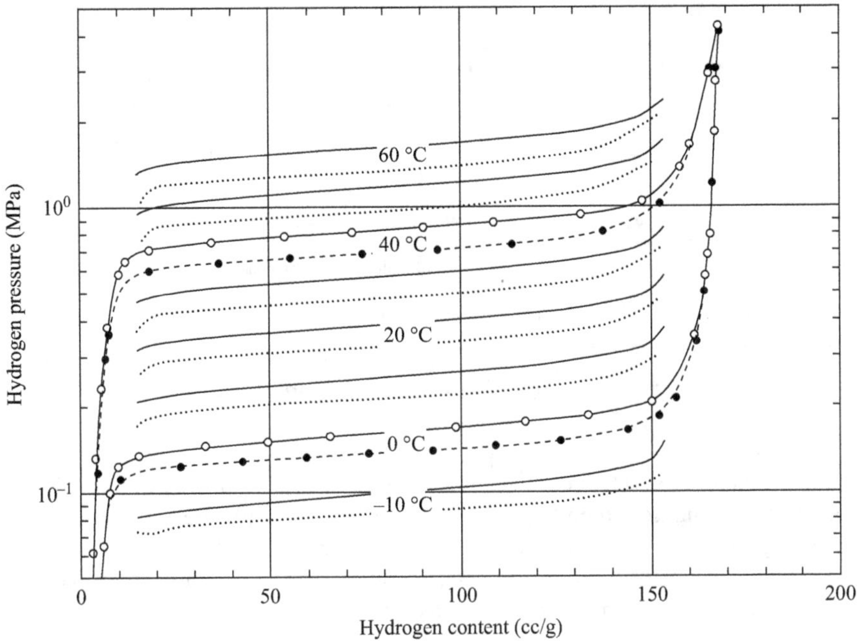

Figure 6.5 Pressure-content-temperature (PCT) diagram of AS5-type metal hydride

the ocean, the functional system architecture for various functional evaluations/verifications is clarified, and they ensure safety to prevent loss of the vehicle during sea trials. The control system was developed with the following principles:

- Both software and hardware are modularized into separate functions in order to prevent problems from spreading.
- Hierarchical architecture is adopted for decision-making based on a wide variety of information from external sources. The received information is organized at the lower level and transmitted to the upper level. At the upper level, the desired value of manoeuvring control is defined from information including navigation plans and observed values.
- To ensure safety, each control system/control logic has the capability to make independent decisions about a stop/emergency surfacing in order to prevent runaway or loss of the vehicle during long-distance cruising.

The hierarchical architecture for decision-making is composed of three levels, as shown in Figure 6.6: mission level, planning level, and execution level. The respective functions consist of mission level, planning level, and execution level.

Mission level

This level integrates the control of an AUV, including controlling the control modes such as starting, stopping, and changing autonomous/remote control, and managing the execution commands in autonomous control. It also provides initial data setting at start-up, constant condition monitoring, failure detection, and failure

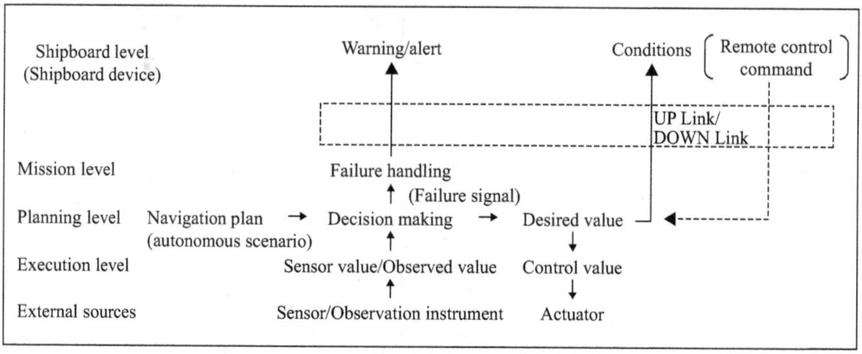

Figure 6.6 Signal flow for autonomous control

handling. The mission level consists of an integrated control process, initial control process, and condition monitoring process to automatically perform tasks such as starting, stopping, and changing autonomous/remote control, managing the navigation plans, setting the initial data, monitoring the conditions, and handling failures. The details of each process are as follows:

- **Integrated control process**: Control the control modes, manage time, and monitor autonomous control.
- **Initial control process**: Power supply control and initial data setting.
- **Condition monitoring process**: Constant condition monitoring and failure detection in any control mode, and failure handling in case of failure.

Planning level

This level determines the execution commands and navigation plans (autonomous scenarios) entered from the shipboard device into detailed execution commands classified by the type of controller. It also collects navigation information such as sensor values and observed values from the execution level, manages the execution of the developed execution commands, and defines the desired value of manoeuvring control. The planning level consists of the navigation information process, navigation pattern execution control process, remote control process, manoeuvring control process, and observation control process. The details of each process are as follows:

- **Navigation information process**: processes navigation signals such as sensor values received from the execution level; determines the position of the vehicle with information from several position sensors, and includes a data analysis logger and receives data from the shipboard device.
- **Navigation pattern execution control process**: manages the execution of navigation/observation instructions in autonomous control.
- **Remote control process**: executes the remote commands received from the shipboard device.
- **Manoeuvring control process**: performs manoeuvring control.
- **Observation control process**: manages the execution of observation commands.

Execution level

This level actuates the controllers given from the planning level, such as actuators, sensors, power supply, and observation instruments, and sends the sensor values, observed values, and operating conditions of each actuator to the planning level. The execution level consists of a sensor control process, actuator control process, power supply control process, communication control process, and observation control process in order to start/control respective equipment. As described earlier, the three hierarchical levels of decision-making enable clear, stable, and safe decision-making at the upper mission level. The hierarchical architecture can provide robust and safe autonomous cruising, which is also easily evaluated and verified.

6.1.3 Next-generation AUV

The author developed the long cruising autonomous underwater vehicle Urashima and established a world record for the longest autonomous cruising (317 km) in February 2005. Recently, development of a longer distance cruising underwater vehicle is required for wide-range sea exploration. This chapter describes the further research of an underwater vehicle for longer distance cruising than Urashima in Japan.

6.1.4 Design process

The author has started to develop the next-generation AUV. Figure 6.7 shows the ideal method devised in this project for developing underwater vehicles. As shown in the flowchart on the left side of Figure 6.7, the underwater vehicle development process includes the definition of vehicle specifications, such as configuration, actuator, control, and sensors in accordance with operational requirements; water tank testing of the model for performance verification; and feedback, comparing the verification results with the defined specifications. The important processes, which the author focuses on, are the simulation design process shown in flowchart ① in Figure 6.7 and the model-making process followed by the fluid properties test and manoeuvring control properties test shown in flowchart ②. Model testing through these processes can identify necessary refinements at an early stage, that is, before construction of the vehicle, and thereby facilitate the process of basic design, detailed design, and construction of the vehicle afterwards. By adopting this method, the risk of reconstruction and major modifications after construction of the vehicle can be reduced to minimize the overall development risk.

Figure 6.8 shows a layout of the vehicle proposed in view of the following challenges. The key technologies are the optimization of vehicle configuration using computer fluid dynamics (CFD) and the control system to ensure fail-safe operation. The vehicle has an X-shaped wing (tail) as shown in Figure 6.9. The X-shaped tail enables it to have a larger wing surface, compared with the cross-shaped tail shown in Figure 6.9 ①, when the underwater vehicle is stored in a shipping container. In addition, it features high capacity lift, high lateral force, high manoeuvrability, smooth berthing and storage, and easy system backup in case of wing failure. The X-shaped tail of the developed model can be changed to a conventional cross-shaped tail and vice versa. The control of the X-shaped tail is based on a wing force distribution

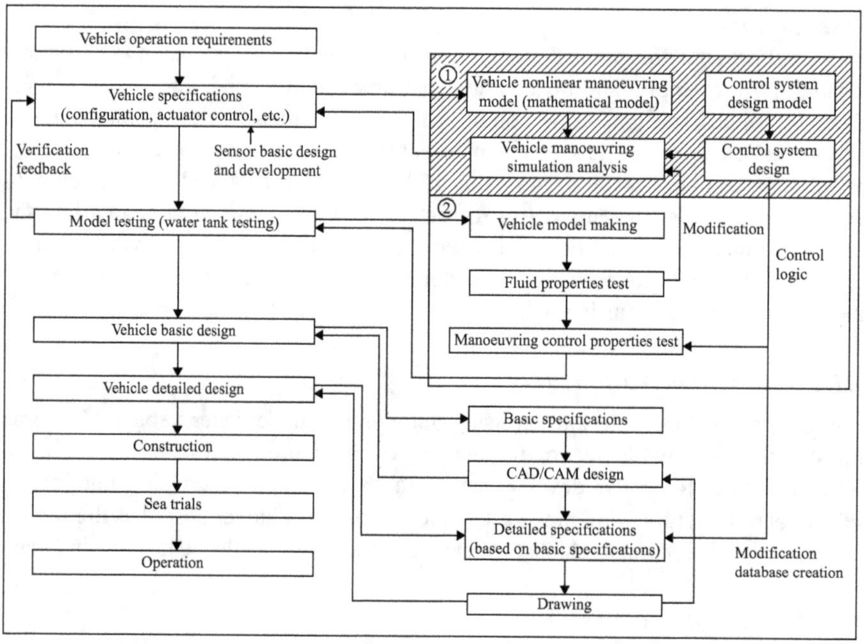

Figure 6.7 Design process for developing underwater vehicles

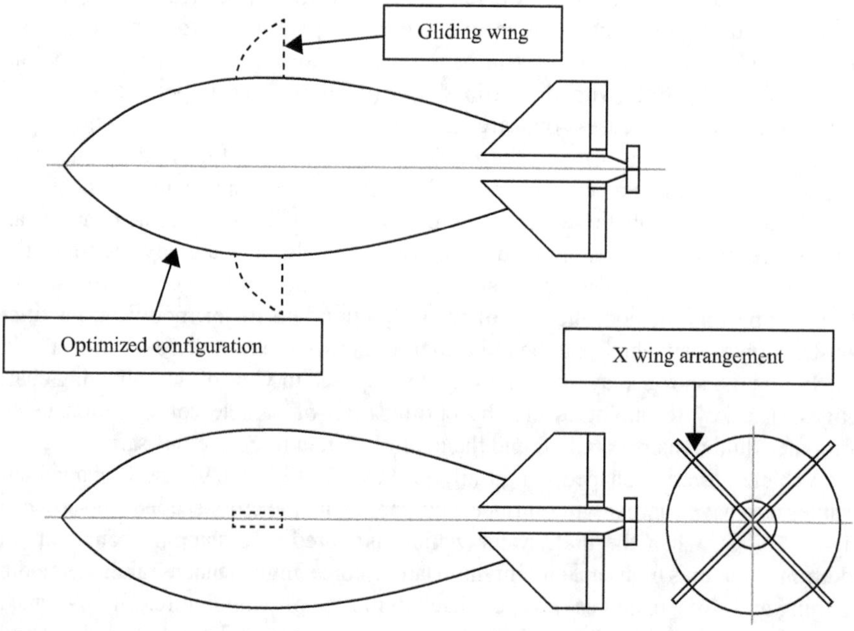

Figure 6.8 Layout of new underwater vehicle

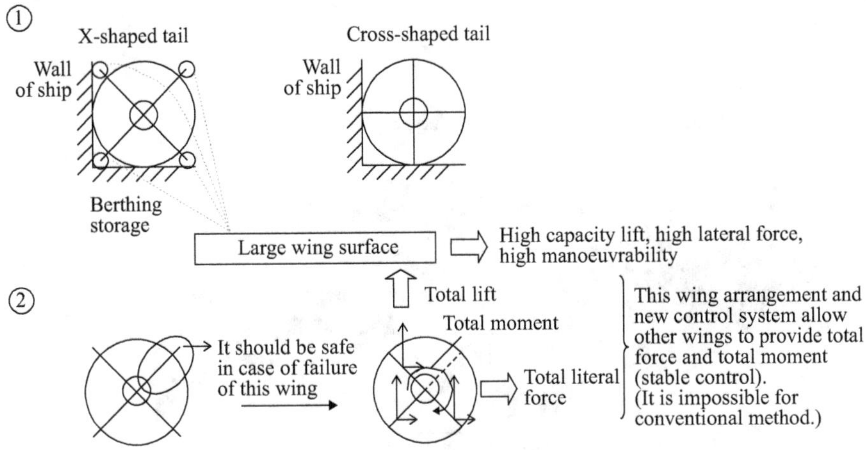

Figure 6.9 Features of the X-shaped tail

method by non-linear programming of the penalty function law for consideration of wing angle limitation, and the semi-Newtonian law to minimize the energy function of total forces and moments in six degrees-of-freedom motion. This wing arrangement and new control system enable the wings to provide total force and moment for stable control, when one wing is broken during operations, as shown in Figure 6.9 ②.

6.1.5 Design and manufacturing

The low-resistance underwater vehicle configuration was simulated using CFD on the basis of the Reynolds-averaged Navier-Stokes equations. The simulation results are shown in Figure 6.10. In Figure 6.10 on the left is a conventional cylinder-shaped vehicle, and on the right is a new configuration designed to provide uniform surface pressure distribution indicated by the same colour. Reduced pressure fluctuation results in reduced flow turbulence and hence minimum drag force. Figure 6.11 shows the appearance of the model constructed from the simulation results, and Figure 6.12 shows its interior. The model is 2.6 m in length, 0.5 m in width, and 0.4 m in height. It is equipped with a lithium battery, ballast tank, motion sensor, and control system to operate autonomously without cables. The control software was developed in the C language. All data is logged and reflected clearly the minimized total wing force (sum squared).

A design method using the above-mentioned CFD method was also developed to design a vehicle from computer-aided design (CAD) data used for the model construction. Figure 6.13 shows the model designed by the design method, and Figure 6.14 shows a model designed by computer graphics to have hydrofoils for higher lift. On the basis of the computer graphics data, the model was constructed as shown in Figure 6.15, which is a glider-type underwater vehicle. The underwater vehicle can cruise like a glider in the water. The manoeuvrability of the conventional cylinder-shaped model and the lower resistance underwater vehicle

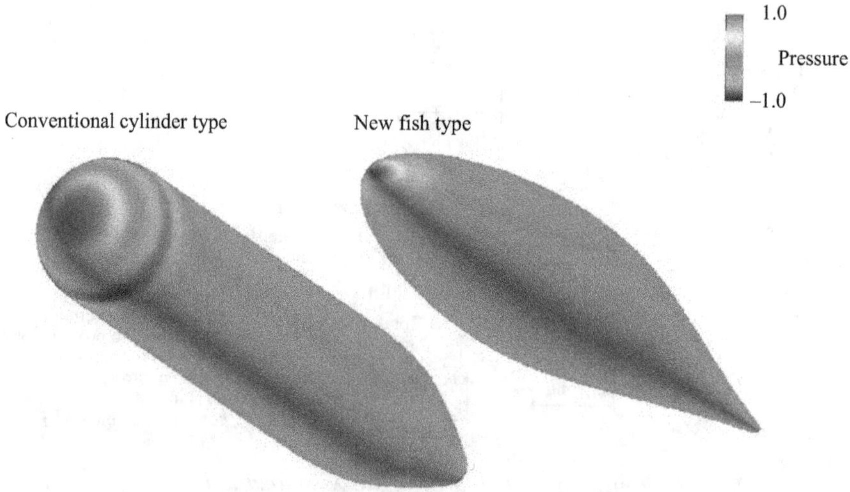

Figure 6.10 Optimum vehicle configuration from CFD simulation

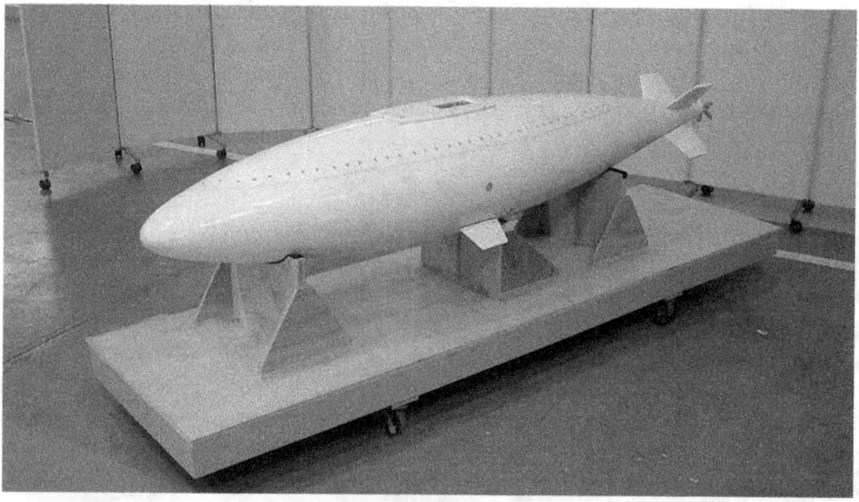

Figure 6.11 Model of optimum vehicle configuration

configuration model (with/without hydrofoils) were verified by water tank testing. The test results are shown in Figure 6.16.

6.1.6 Tank test

The water tank testing and analysis indicates that the next-generation model has a considerably longer cruising distance than the conventional (cylinder-shaped) type because of the glider effect of the hydrofoils and less drag force, as shown in Figure 6.16, which is evaluated by maintaining a constant vehicle volume and

Figure 6.12 Model with internal structure

Figure 6.13 Model design by computer graphics

battery capacity. In addition, the system allows for stable control with at least one wing failure. The other wings can compensate for cruising of the vehicle using the control software. The glider cruising has the advantage of extending the cruising distance, especially using its own lift forces. Additionally, if the vehicle operates in deeper sea, it can cruise longer depending on the depth.

In addition, the X-shaped tail was tested and is especially effective for the zig-zag cruising test shown in Figure 6.17. Higher manoeuvrability was verified by the tank test.

Figure 6.14 Underwater vehicle with hydrofoils designed by computer graphics

Figure 6.15 Model with hydrofoils

6.1.7 Conclusion

The author developed elemental technology for long-distance cruising of an autonomous underwater vehicle for sea exploration. This is in the process of production of the next-generation marine observation system network in Figure 6.18, combined with other developed ROVs and aviation robotics. In addition, the ROV has a compact portable observations system shown in Figure 6.19, which enables

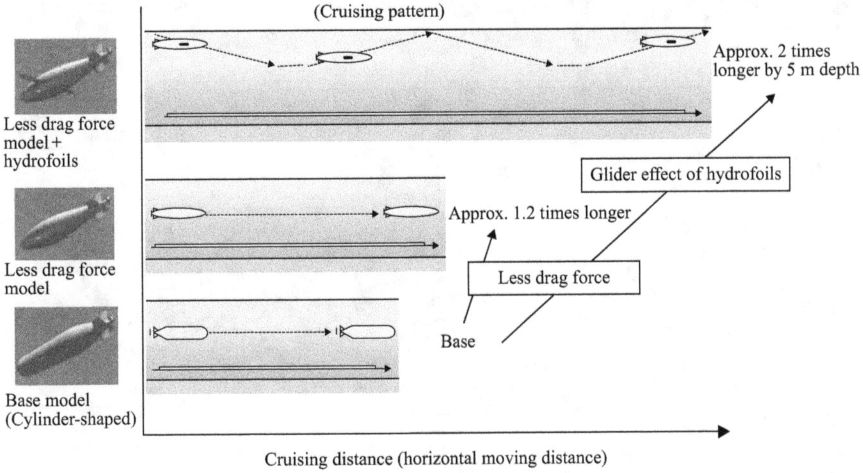

Figure 6.16 Comparison of cruising range between next-generation and conventional model

Figure 6.17 Zig-zag cruising test

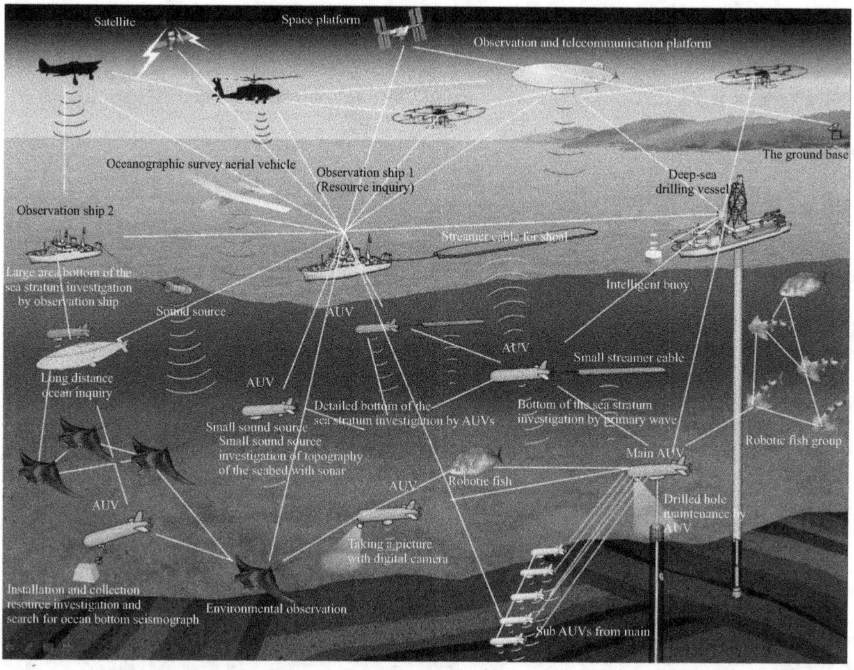

Figure 6.18 Concept of a next-generation marine observation system network

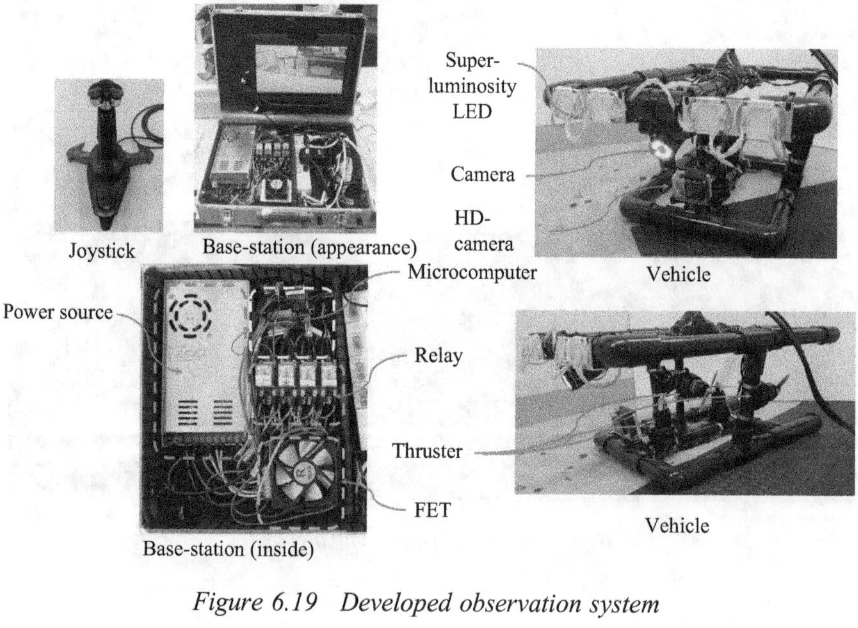

Figure 6.19 Developed observation system

the detailed sea observation after the AUV's macro exploration in the ocean. Finally, the author would like to express his sincere gratitude for JAMSTEC and Mitsubishi Heavy Industries, Ltd. for their cooperation.

6.2 Remotely operated vehicle

6.2.1 Introduction

The aging structure of ports and dams needs to be checked for diagnosis of degradation. In addition, offshore structures are increasing, especially for renewable energy, and undersea maintenance systems are required for checking for cracks. In addition, there is a need to develop practical unmanned underwater observation systems for disaster prevention. The observation system must accomplish the purposes of seabed monitoring to evaluate damage after a disaster and evaluate structural objects in ports, at dams, and at offshore facilities. They must feature high mobility, low-cost, and safe operation.

In general, the underwater observation of facilities in offshore areas such as a harbour or fish farm is conducted manually by divers. However, observation by divers has some difficulties such as high cost, danger to workers, and excessive time needed for expansive observation. Although many research results for ROVs have been reported, it is often the case that the system is large and expensive.

There is therefore a growing need for small and low-cost unmanned underwater observation vehicles with highly manoeuvrable control systems.

6.2.2 Concept of underwater vehicles

The author and his colleagues developed the vehicle with consideration for mobility, controllability, transportability, and practicality. The concept of the proposed underwater vehicle is as follows:

- System is small size (619 mm L × 419 mm W × 233 mm H, 5.6 kg)
- Remotely controlled with a motor power cable
- Start-up is within 3 min for experiment
- Simple controllability
- Unmanned observation system
- Video image is monitored at base station
- Operating time is over 2 h per operation
- High-powered light
- Low cost

Figure 6.20 shows the developed ROV to satisfy the concept. The proposed system was chosen specifically to target a depth range of 0–100 m.

6.2.2.1 Frame design with CAD

The frame of the ROV is first designed using CAD. Table 6.5 shows the specifications of the proposed vehicle. Figure 6.21 illustrates a CAD design of the body of the ROV. The vehicle body is designed as a unique structure based on the 'principle

Figure 6.20 Developed ROV

Table 6.5 Specifications of the proposed vehicle

Body	Length	619	mm
	Width	419	mm
	Height	233	mm
	Weight	5.6	kg
Material	Polyvinyl chloride pipe		
Cable	Length	0–100	m
Motor	Mayfair marine motor cartridge × 5 units		
	Current	2–4	A
	Voltage	48	V
Camera	GoPro HD HERO2 (Full HD)		
	Resolution	1920 × 1080	px
	Frame rate	30	fps
	View angle	170	degrees
Light	Super-Luminosity LED array × 4 units		
	Output power	max 1200 per LED array	W
	Current	2	A
	Voltage	38	V/unit
	Light flux	14400	lm/unit
Speed		0.5~1.0	m/s

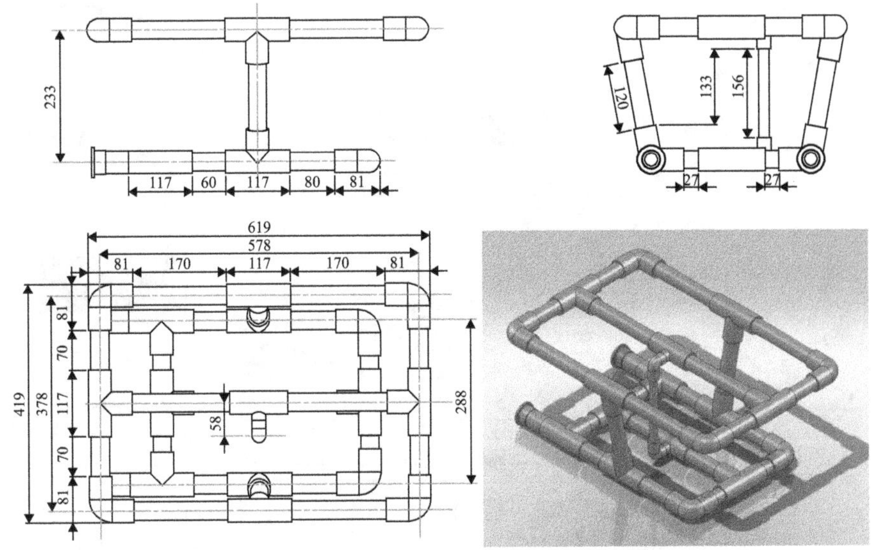

Figure 6.21 Frame design of the vehicle

of float and weight' (licensed by the author and his colleagues). The upper frame operates as a float, and the lower frame operates as a weight. The structure provides a well-balanced body without the need for thrusters. The frame is made of an impact-resistant polyvinyl chloride (PVC) pipe. A moving weight adjustment system is manufactured by a 3D printer, and inserted inside each lower long pipe.

6.2.2.2 System function design

The developed observation system has the following features:

- Power source: 900 W gas engine generator
- Light: Super-luminosity LED array (300 W, 14400 lm) × four units
- Camera: Full HD camera (1920 × 1080 px, 30 fps, 170° of view)
- Controller: Control support system by microcomputer

Figure 6.22 illustrates the system configuration diagram of the ROV. A microcomputer is utilized as a controller, and the control support intelligence system is installed in the computer. A proportional control system signal is sent to the motor controller through the microcomputer. Table 6.5 shows the specifications of the proposed vehicle, and Table 6.6 shows the specifications of the proposed observation system. In order to operate the observation system anywhere, the GAS engine generator was adopted as the power source. An electronic circuit, monitor, and recorder are embedded in the trunk case (base-station) for portability. The vehicle mounts four super-luminosity LED arrays. The brightness of the LEDs is controlled by a light control circuit. Figure 6.23 shows a real-time picture for monitoring of a structure in the sea of Nagasaki, Japan.

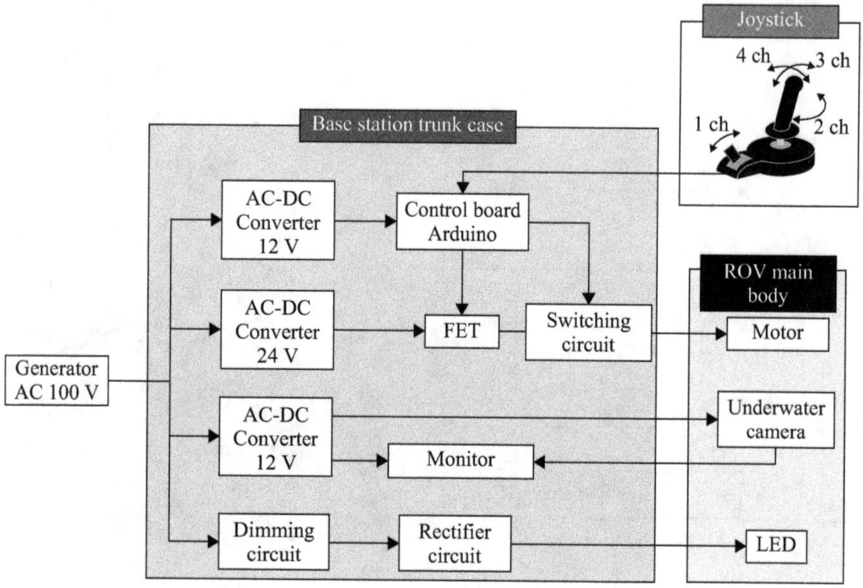

Figure 6.22 ROV system configuration

Table 6.6 Specifications of the proposed observation system

System configuration	ROV, cable, base-station (trunk), power generator
Power source	900 W GAS engine generator
Operating time	2 h (2 gas cartridges)
Target water depth	0–100 m
Monitoring features	Real-time monitoring, Video recording, High-powered light (Super-Luminosity LED)
Operation procedure	Proportional control system
Motion	Forward, back, side, rotating, surfacing, diving
Start-up	Within 3 min
Minimal operators	2 persons

6.2.3 Observation system

Figure 6.19 shows the developed observation system. Each element of the observation system is shown as follows:

6.2.3.1 Wire cable

A signal wire and power wire are bundled together in a wire cable. The wire cable has the features of being waterproof, pressure-tight, and flexible. A floating buoyant material is set at about 1-m intervals to avoid impact on the vehicle as shown in Figure 6.24.

Figure 6.23 Monitoring of a structure in the sea

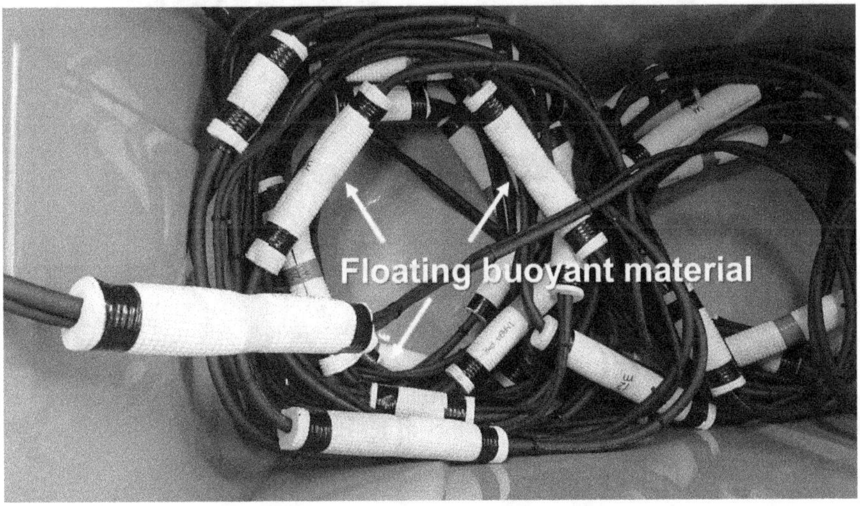

Figure 6.24 Floating buoyant materials

6.2.3.2 Portable base station

The base-station consists of the electronic circuit, monitor, and recorder. The operator is able to operate the vehicle while watching the screen as shown in Figure 6.25.

6.2.3.3 Control system

The operator controls the vehicle with the proportional control system. The control method of the motor system is pulse width modulation (PWM) for power control.

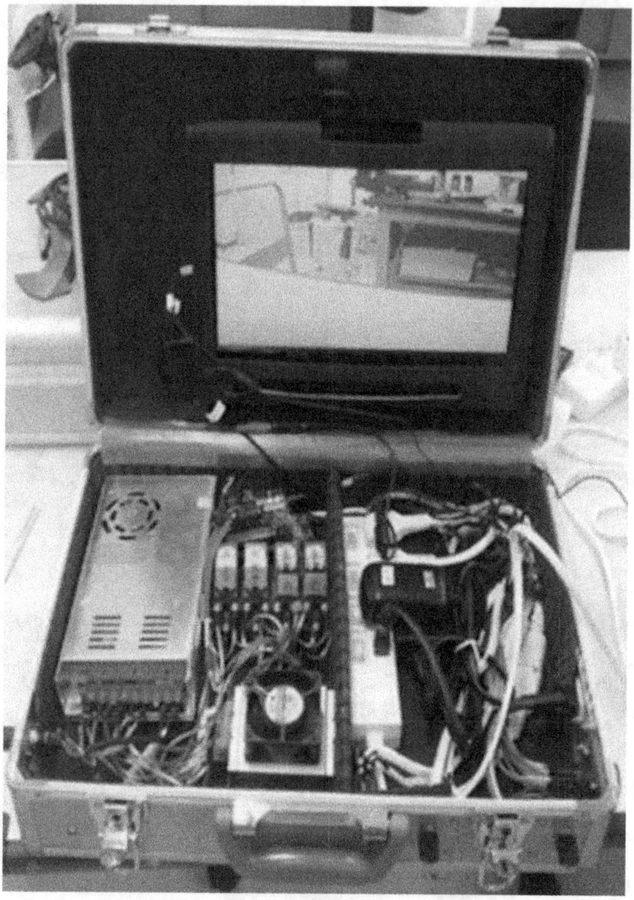

Figure 6.25 Portable base station

The signals from the joystick are converted automatically into direction information signals, which are sent to five motors using PWM. This task is carried out by the microcomputer.

6.2.4 Experiments and results

The author and his colleagues conducted experiments at Okinawa Chatan offshore (Okinawa, Japan) and verified the practical effectiveness of the developed unmanned underwater vehicle (Figure 6.26). The vehicle was put in the sea from the beach, and observed the targets in the sea. The experiments were conducted in the open sea in the presence of waves and tidal currents. The main purposes of these experiments were verification of video images within a limited time, and controllability of the vehicle under real sea conditions. The author and his colleagues could observe and record moving images of targets within the limited time (10 min) as shown in Figures 6.27–6.29.

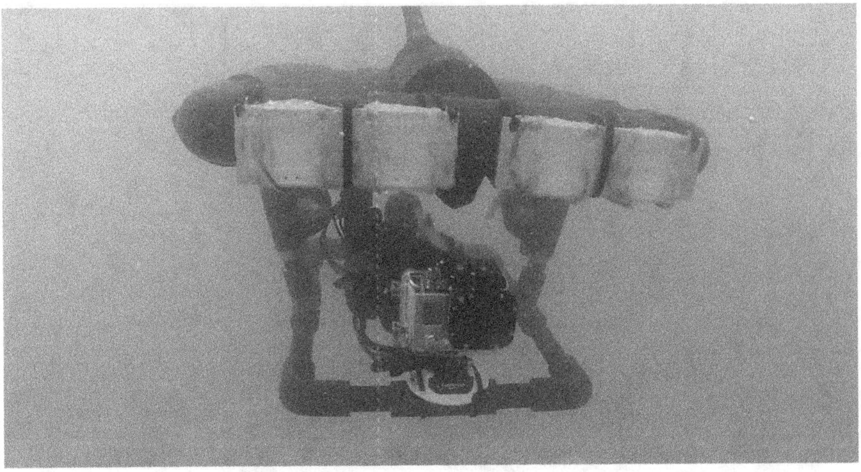

Figure 6.26 ROV experiments in the sea

Figure 6.27 First target of observation in the sea

It was somewhat difficult to control the vehicle owing to the effects of the strong tidal current (1.0–1.5 m/s) and the wind speed (10 m/s). However, the system is easy to operate under severe environmental circumstances.

The author and his colleagues won the best ROV award in the Underwater Robot Contest Okinawa, Japan, held on December 6, 2014. The performance of the ROV is verified through the sea trials and winning the best ROV award. In addition,

Figure 6.28 Second target of observation in the sea

Figure 6.29 Last target of observation in the sea

the ROV has been successfully tested for inspection of a dam and observation of a fish farm in Japan.

6.2.5 Performance evaluation

The developed underwater observation vehicle was designed based on on-site needs. Thus the author and his colleagues developed the vehicle with considerations of mobility, controllability, transportability, and practicality. The results of performance evaluations of the developed vehicle are described as follows:

6.2.5.1 Mobility

In the sea, in the presence of waves and tidal currents, the vehicle was able to exhibit high mobility by improvement of the ROV structure from the Shibushi Offshore sea trial experience in Kagoshima, Japan, 2013.

6.2.5.2 Controllability

The vehicle was controlled by a proportional control system. Consequently the operator was able to operate the underwater observation vehicle without specialized experience or advanced techniques. The operation of the vehicle while viewing a monitor on the base station is somewhat difficult to orientate. It is necessary that the operator should take some time to practice operating the vehicle. In addition, side slide movement is possible using the side propeller.

6.2.5.3 Transportability

Because the developed vehicle is designed to be small-scale and lightweight, the vehicle is easy to carry around and is loaded into a normal automobile. Furthermore, the vehicle system is able to start operating within 3 min of arrival at the observation field.

6.2.5.4 Practicality

The results of the experiments showed that the vehicle is amply practical in the sea and for use in deep dams. However, more robustness and ruggedness of the vehicle systems are required, because the vehicle is thrown into the sea, raised from the sea, and operated under the strong tidal currents in Goto Island, Nagasaki, Japan.

6.2.6 Conclusion

In this research, the author and his colleagues developed a high-mobility unmanned underwater observation vehicle for sea exploration, and conducted experiments in the sea and in a deep dam. The results of the experiments show that the unmanned underwater observation vehicle is amply practical for the desired use. The developed underwater vehicle features high-mobility design (small scale, lightweight, and low cost), an underwater observation system (compact size underwater camera, LED light, wired transfer unit, and base station monitor), a remote-control system (embedded microcomputer and proportional control system), and a high-powered light (super-luminosity LED array).

An advanced underwater vehicle is in the process of production by the author and his colleagues on the basis of the experimental results. Hereafter, the author and his colleagues plan to advance the further development of the vehicles by adding the developed ROVs as shown in Figure 6.30.

6.3 Offshore robotics and mechatronics

6.3.1 Introduction

This section describes the development of a new robust dynamic positioning system for an offshore oil development platform. The system is developed by thruster force

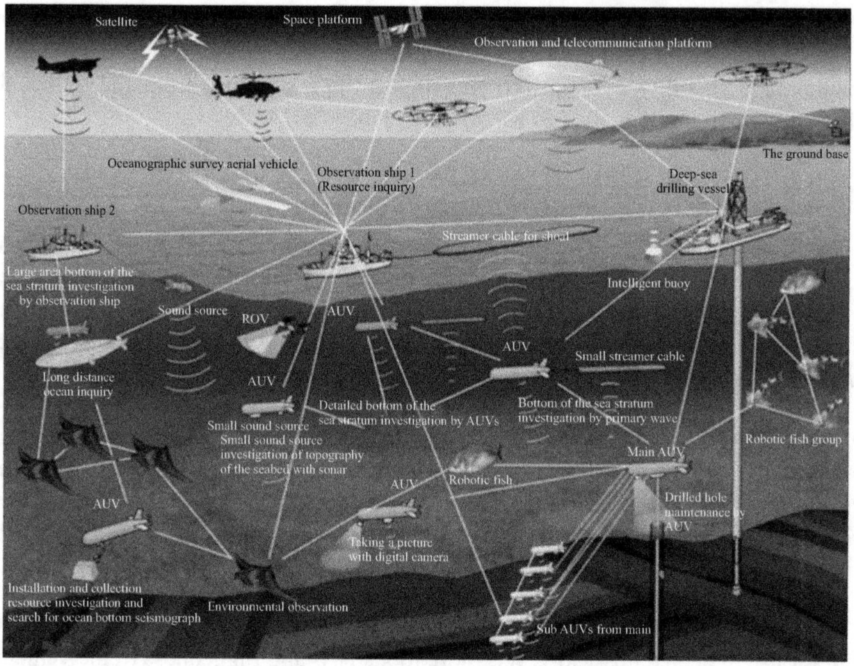

Figure 6.30 Concept of a next-generation marine observation system network using ROVs

distribution control, and the system robustness against actuator failure is much improved; full operation against severe environmental conditions can be executed. The effectiveness of the system was confirmed by computer simulations and real experiments.

The development of new oil-bearing areas has been widely expected to address the energy shortage problem in the world. Thus, new offshore oil-bearing areas need to be developed; however, the number of them has been decreasing year by year. Therefore, the development of new oil resources at sea is moving gradually to deeper sea areas, that is, from existing continental shelves to areas where the continental shelf slopes down to deeper waters. Further, these days it can be expected that oil resource development in deep-sea areas will experience much more severe environmental conditions. In this situation, the floating production system is receiving more attention as a system suitable for marginal oil development. Figure 6.31 shows the constituent elements of a floating production system. As small-scale marine oil development takes place in deeper sea areas, position-keeping has become more difficult, when using only existing mooring methods. This is because of technical and financial aspects. It is therefore necessary to apply a dynamic positioning system, hereinafter referred to as DPS, in conjunction with thrusters. In order to assess the performance of platform control with DPS that meets its specification requirements, DPS logic control development and behaviour analysis simulation tools are needed.

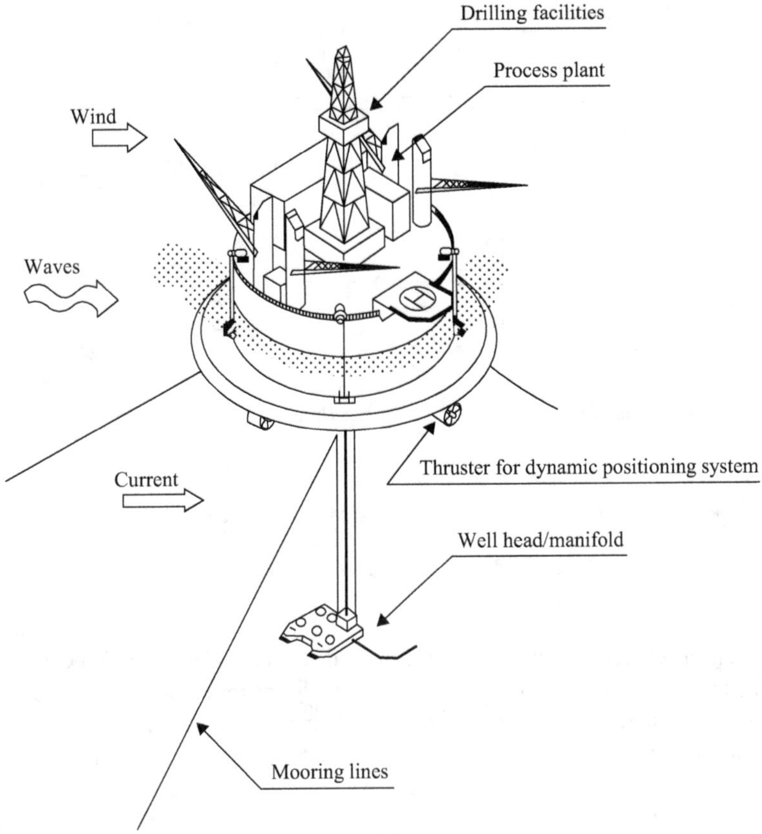

Figure 6.31 Example of a floating production system

The author has developed DPS control logic based on a nonlinear programming method, and verified the efficiency of the control logic through real-time control tests at a tank experimental facility. At the same time, the author and his colleagues have developed behaviour analysis simulation technologies for a platform with DPS and established a technological assessment of the platform with DPS in 2001. The fault tolerance design of the system is a focus of this section.

6.3.2 Control algorithm

The control logic of DPS is based on a nonlinear programming method developed by the author from 1991 to 2001.

In most previous cases, the control algorithm was restricted by the type of platform, the number and type of actuators, and the system's nonlinear characteristics. However, the control logic has reduced those restrictions in order to widen the logic's applicability. As shown in Figure 6.32, the basic control logic consists of a thruster force distribution element, a platform movement element, a command generation element, and a thruster manipulation signal determination element.

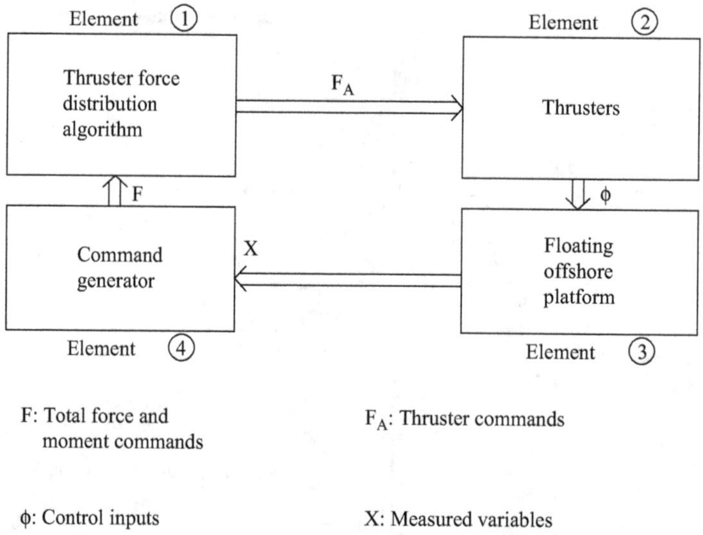

Figure 6.32 Basic control logic of DPS

The thruster force distribution element distributes the total force and total moment to be established on each thruster. An algorithm of the thruster force distribution element is based on a nonlinear programming method, and the nonlinear programming method is carried out by the mixed penalty method and the modified quasi-Newton method, in order for it to be applied to marine platform motion, thruster restriction conditions, and mechanical cost functions using nonlinear characteristics. The algorithm is derived by the following five steps:

The control algorithm (five steps)

Step 1: Input thrust value of the thrust initial state and thrust direction.

Step 2: Input mechanical conditions.

$$H_1 = Fxc - \sum_{i=1}^{n} Fxsi \tag{6.1}$$

$$H_2 = Fyc - \sum_{i=1}^{n} Fysi \tag{6.2}$$

$$H_3 = Nzc - \sum_{i=1}^{n} FxsiYsi - \sum_{i=1}^{n} FysiXsi \tag{6.3}$$

where

Fxc, Fyc : Total force of the body axis fixed coordinate system in the x, y direction
Nzc : Total moment around the body axis fixed coordinate system z

$Fxsi, Fysi$: Body axis fixed coordinate system x, y direction component force of the nth thruster

Xsi, Ysi : Body axis fixed coordinate system x, y direction fitting position of the nth thruster

n : The number of thrusters

Step 3: Input actuator control condition. Use thruster azimuth for the actuator and set the control condition as follows:

$$G_1 = F^2 \max si - (F^2 xsi + F^2 ysi) \tag{6.4}$$

$$G_2 = (F^2 xsi + F^2 ysi) - F^2 \min si \tag{6.5}$$

Step 4: Input the cost function, which is the square sum of the following thrust and minimal energy.

$$f = \sum_{i=1}^{n} F^2 si \tag{6.6}$$

Step 5: Convert the cost function of (6.6) to a total cost function. The total cost function is set as follows, according to the limitations of the mechanical control condition and actuator control condition:

$$F = f + \psi(r, Hi) + \phi(t, Gj) \tag{6.7}$$

ψ of (6.7) should be as follows, according to the mechanical control condition.

$$\psi(r, Hi) = r \sum_{i=1}^{n} Hi \tag{6.8}$$

ϕ should be as follows, according to the actuator control condition.

$$\phi(t, Gj) = t \sum_{j=1}^{n} \frac{1}{Gj} \tag{6.9}$$

Here, r and t are parameters.

The optimum thrust distribution is found by Step 1 through to Step 5. In addition, the optimum position will also be found even if the thruster position is variable. After determining the thrust distribution, the thruster positioning element command value with the thruster manipulation signal determination element is calculated. Consideration of the thruster fixed characteristics should be included.

Total force and total moment to the thruster force distribution element is formulated from a command generation element. On the basis of a deviation signal from the platform position coordinates, position target and azimuth target are set; the proportional gain, differential gain, and integration gain are multiplied; and the total force and moment for keeping a fixed position are calculated.

The total force and total moment to the thruster force distribution element, which are calculated from the current and wind disturbance model, and a built-in feedforward element, as a prior control signal, can easily be bias-added to the command generation element in order to input to the thruster force distribution element. In the same manner, it is possible to expand various functions such as joystick control and unknown presumed disturbance estimation controls.

6.3.3 Computer simulation

The simulator program consists of input data, a data set component, fluid force calculation component, control component, movement analysis component, and output component, as shown in Figure 6.33.

Force and moment by current are calculated from (6.10) to (6.12).

$$F_{XC} = 1/2\rho\nabla^{2/3}U_C^2 C_{FX} \tag{6.10}$$

$$F_{YC} = 1/2\rho\nabla^{2/3}U_C^2 C_{FX} \tag{6.11}$$

$$N_{ZC} = 1/2\rho\nabla^{2/3}L_O U_O^2 C_{MZ} \tag{6.12}$$

where

ρ : Density of seawater
∇ : Displacement of platform
L_O : Length of lower hull
U_O : Velocity of current
C_{FX}, C_{FY}, C_{MZ} : Coefficients of current force

Force and moment of the wind axis are calculated from (6.13) to (6.15).

$$F_P = 1/2\rho_a A U_W^2 C_{FP} \tag{6.13}$$

$$F_Q = 1/2\rho_a A U_W^2 C_{FQ} \tag{6.14}$$

$$M_R = 1/2\rho_a A L_O U_W^2 C_{MR} \tag{6.15}$$

where

ρ : Density of atmosphere
A : Sectional area of platform
L_O : Length of lower hull
U_W : Velocity of current
$C_{FP}, C_{FQ}, C_{FR}, C_{FMP}, C_{MQ}, C_{MR}$: Coefficients of wind force (wind axis)

Force and moment of the body axis are given by (6.16–6.18).

$$F_{XW} = F_P \cos\beta_W - F_Q \sin\beta_W \tag{6.16}$$

$$F_{YW} = F_P \cos\beta_W + F_Q \cos\beta_W \tag{6.17}$$

$$M_{ZW} = M_R \tag{6.18}$$

where

β_W : Wind direction

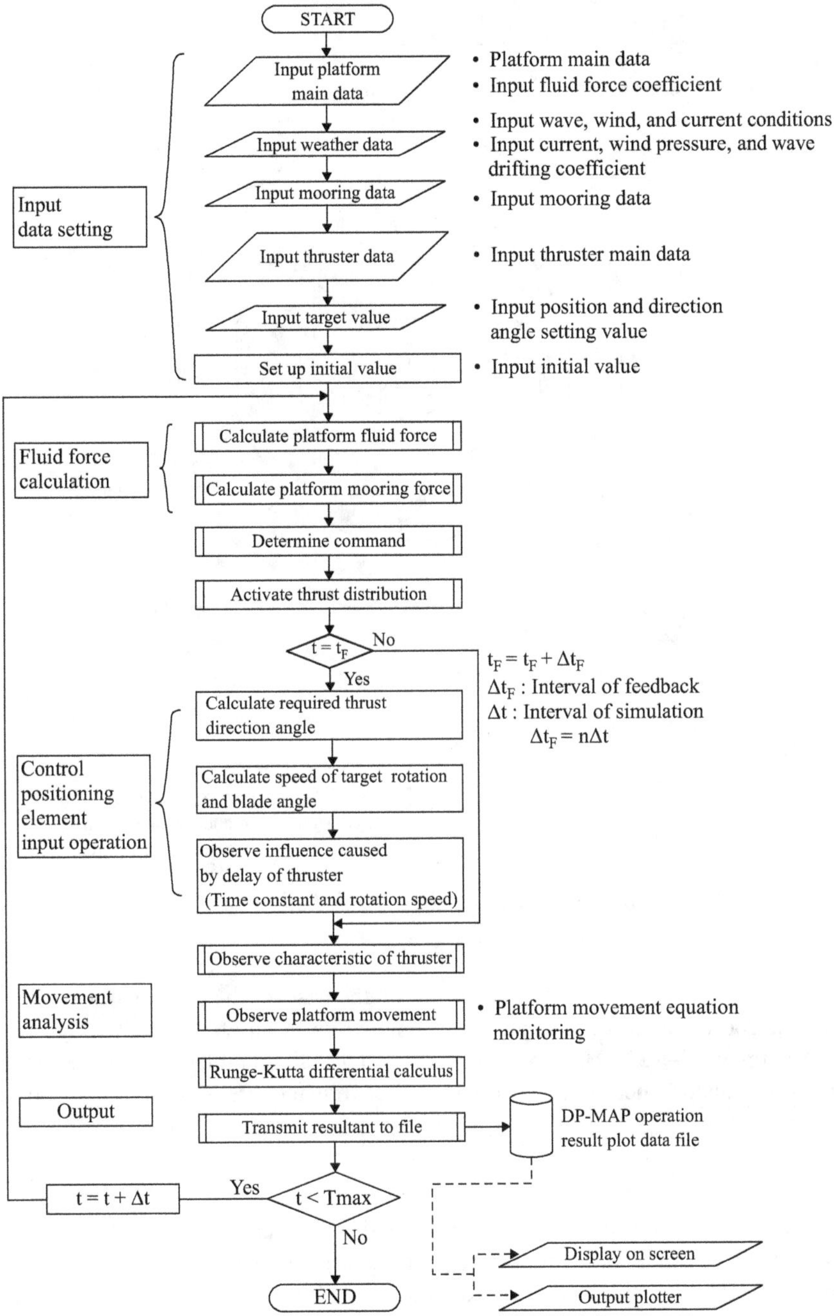

Figure 6.33 Flow chart of basic DPS simulation program

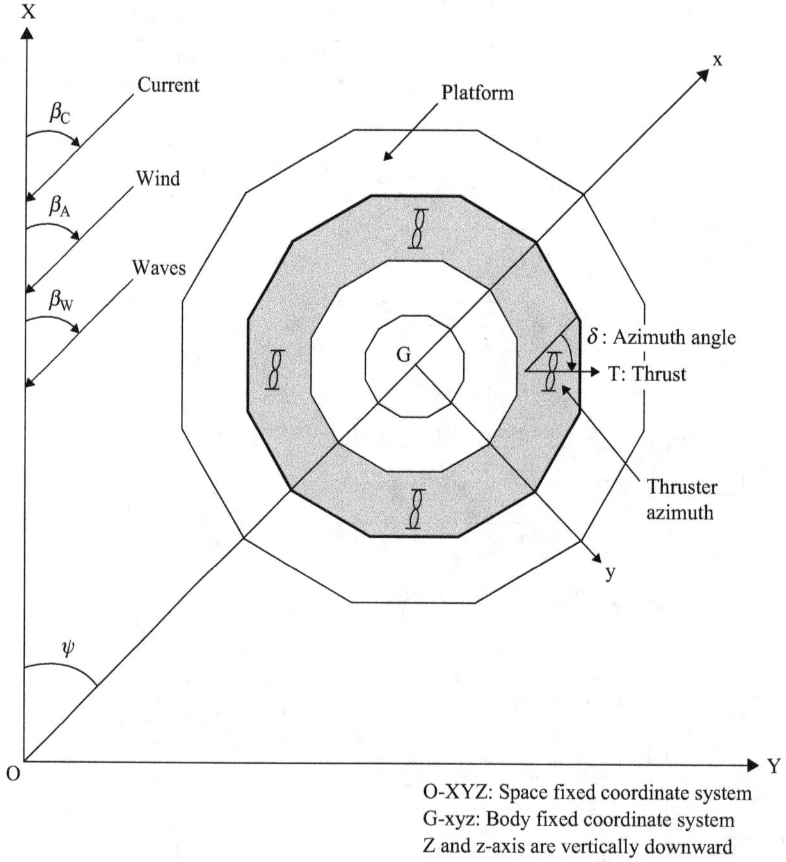

Figure 6.34 Coordinate system of platform with DPS

Wave force and moment are calculated on the basis of the estimated wave data. Also, the coordinate system of platform is shown in Figure 6.34.

Figures 6.35–6.37 show examples of the DPS simulation under wave, wind, and current disturbance in the case where four thrusters are located on the mono-column platform.

6.3.4 Experiment

An experiment was carried out to examine the DPS control logic. Figure 6.38 shows the test equipment layout diagram of the offshore platform model. The main components of the platform model are summarized in Table 6.7, and the model test conditions are shown in Table 6.8. Experiments were focused mainly on conditions occurring with high frequency at the ocean platform working sea area. For all the conditions, real-time DPS by thrusters could be performed within 20 m accuracy of the real platform. Examples of the results are shown in Figures 6.39–6.41.

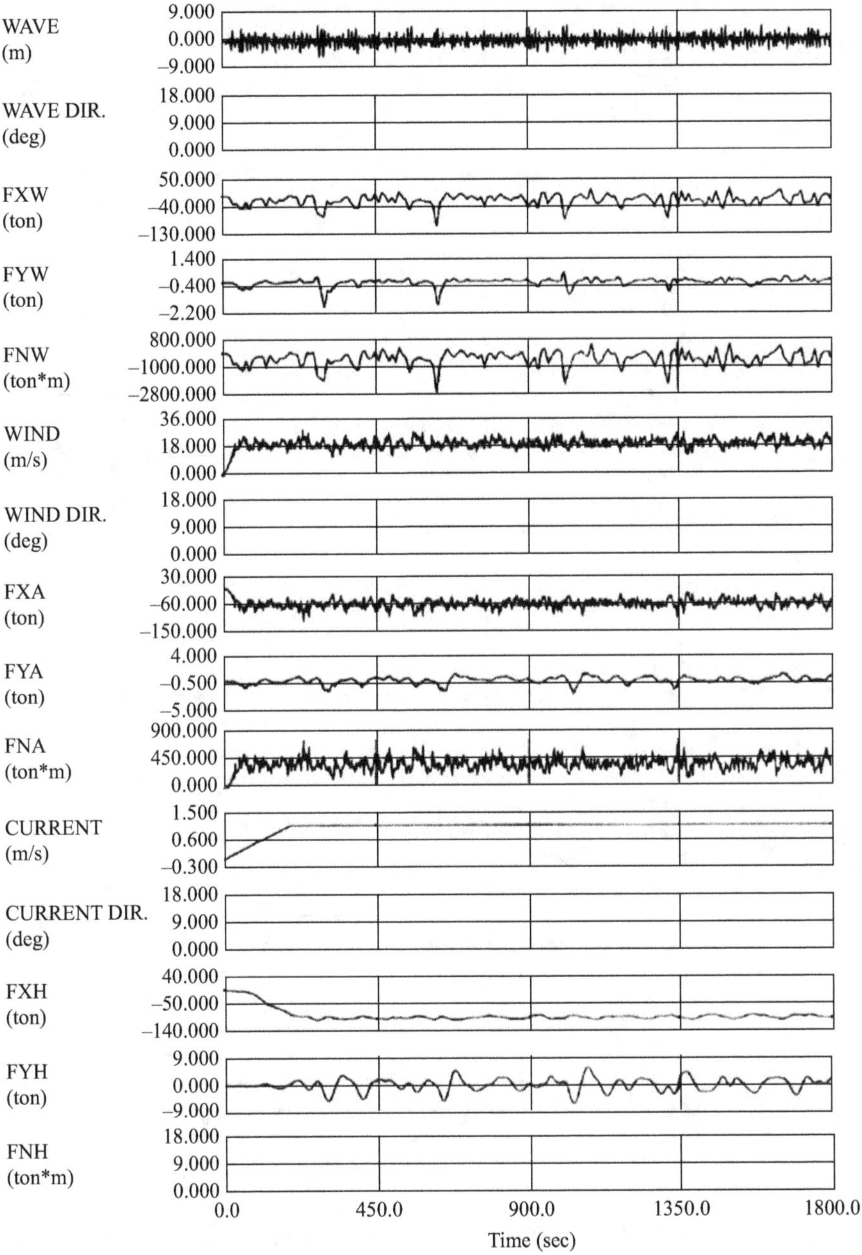

Figure 6.35 Simulated results for DPS (calculation result by simulation/ mono-column-type vessel configuration/4 thrusters)

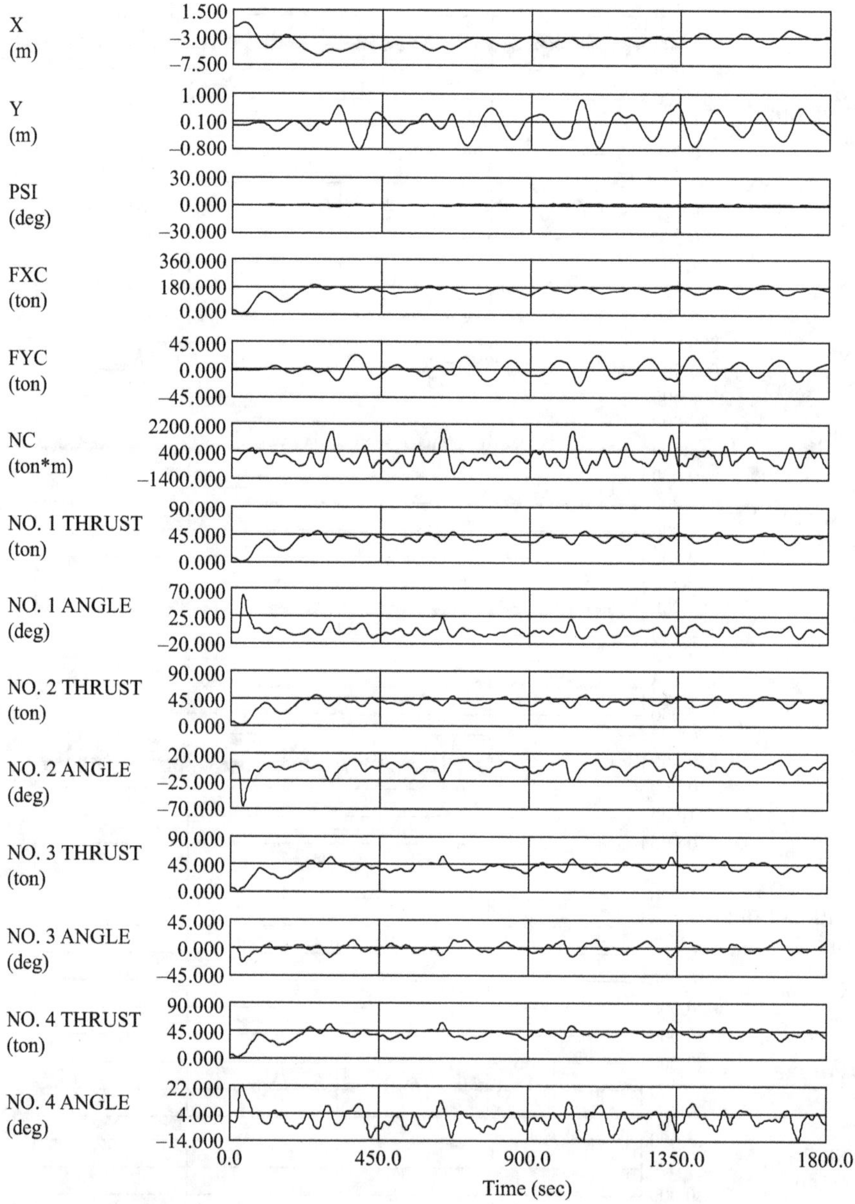

Figure 6.36 Simulated results for DPS (calculation result by simulation/ mono-column-type vessel configuration/4 thrusters)

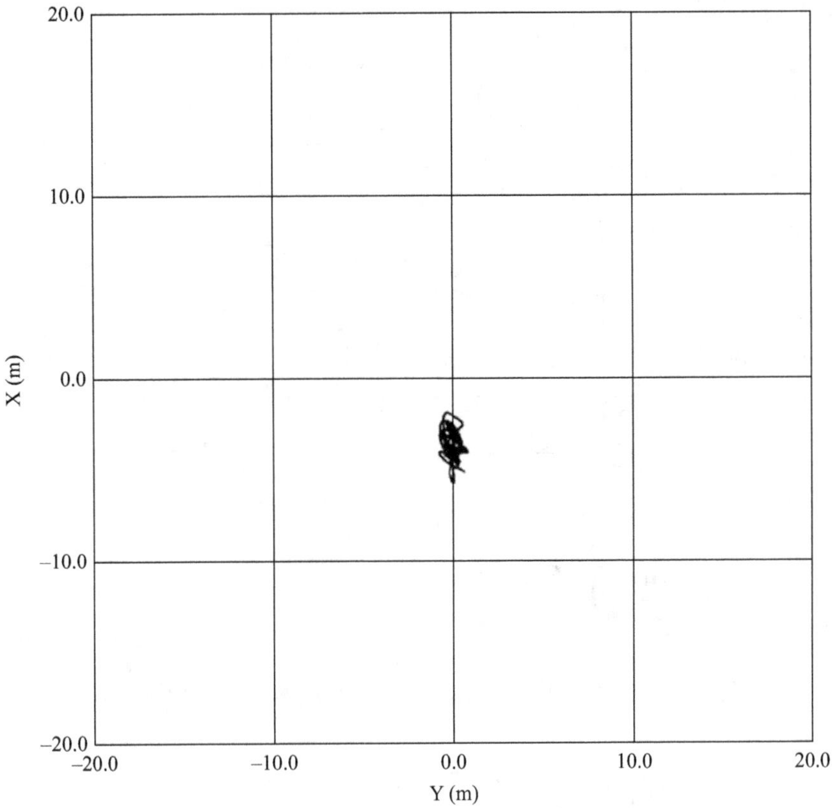

Figure 6.37 Simulated results for DPS (calculation result by simulation/mono-column-type vessel configuration/4 thrusters)

Fault-tolerance testing was conducted on the assumption that one of the thrusters suddenly broke down, while a DPS system with four thrusters was being controlled. However, the DPS system had no problems because the control logic functioned properly, as shown in Figures 6.42 and 6.43.

6.3.5 Conclusion

A new control algorithm based on a nonlinear programming method was developed to improve position-keeping characteristics of an offshore platform. The effectiveness of the control algorithm was confirmed by computer simulations and experiments. Thus, the generality for designing the DPS of a platform and the robustness against failure of an actuator can be much improved by the control algorithm.

Finally, the author and his colleagues are grateful to Japan Oil, Gas and Metals National Corporations, Mitsubishi Heavy Industries, Ltd. for the work conducted through their joint research.

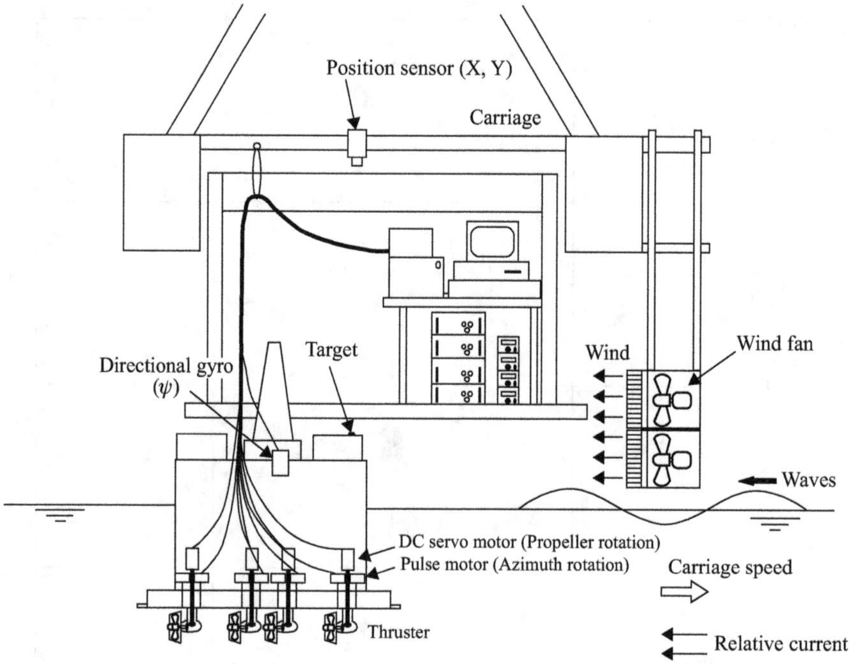

Figure 6.38 Model test equipment layout diagram in the tank

Table 6.7 Main components of platform model, improved mono-column vessel configuration

		Simulator	**Actual model**
Column external diameter	L_{CO}	1.36 m	68.0 m
Column internal diameter	L_{CI}	0.84 m	42.0 m
Well diameter	L_{HI}	0.42 m	21.0 m
Lower hull outer diameter	L_{HO}	1.64 m	82.0 m
Draft	d	500 mm	25.0 m
Lower hull height	D_H	100 mm	5.0 m
Displacement	Δ	530 kg	68,000 ton
Draft area	A_W	0.858 m^2	2145 m^2
Metacentre height	GM	70 mm	3.5 m
Centroid height	KG	380 mm	19.0 m

6.3.6 Updates in DPS: low-cost design method of offshore platform DPS before construction

6.3.6.1 Introduction

To accommodate energy consumption in the near future, new offshore oil-bearing areas have to be developed. These areas are mostly located in deep waters. Position-keeping in deep waters is more difficult, when using only existing mooring methods.

Table 6.8 Test conditions

		Wind (Steady wind)			
		BFT 6	**BFT 7**	**BFT 8**	**BFT 9**
Actual model	Wind velocity	12.6 m/s	15.7 m/s	19.3 m/s	22.6 m/s
	Wind load	23 ton	36 ton	54 ton	74 ton
Simulator	Wind velocity	1.8 m/s	2.2 m/s	2.7 m/s	3.2 m/s
	Wind load	0.18 kg	0.29 kg	0.43 kg	0.58 kg

		Wave (Unsteady wave)			
		BFT 6	**BFT 7**	**BFT 8**	**BFT 9**
Actual model	Significant wave	4.0 m	5.2 m	6.7 m	8.5 m
	Average wave period	7.5 s	8.1 s	9.0 s	10.3 s
	Average wave drift force	33 t	49 t	70 t	79 t
Simulator	Significant wave	80 mm	104 mm	134 mm	170 mm
	Average wave period	1.06 s	1.15 s	1.27 s	1.46 s
	Average wave drift force	0.26 kg	0.38 kg	0.55 kg	0.62 kg

BFT : Beaufort scale

		Current (Steady current)		
		Cur.1	**Cur.2**	**Cur.3**
Actual model	Current velocity	0.8 m/s	1.2 m/s	1.6 m/s
	Current force	50 ton	113 ton	201 ton
Simulator	Current velocity	0.11 m/s	0.17 m/s	0.23 m/s
	Current force	0.37 kg	0.89 kg	1.62 kg

It can be beneficial to use a dynamic positioning system instead of a mooring system. This dynamic positioning system must handle several environmental conditions, from calm water to heavy seas and high wind and current speeds.

The author and his colleagues have developed a new generation dynamic position system after 20 years of research. This system is implemented in a floating production platform model. The simulations show a good improvement of the ship's behaviour. The next step is to carry out hardware functionality tests. It is not practically possible to implement a DPS directly on board an actual ship and start testing it. First, from a safety point of view, this is not preferable. Second, the operation costs per day for a ship are much too high. The first alternative is to perform tests with scaled models. There are several universities and specialized companies which can provide these tests. However, the costs of these kinds of tests are still high. The goal of the initial hardware tests of a DPS is to verify if the system behaves as expected. Initially, influences such as hydromechanical scaling effects and sensor accuracy play a less important role. However, increasing accuracy from these influences makes the research costs much higher.

The author and his colleagues have established a low-cost design method of offshore platform DPS before construction, at one-tenth of cost of conventional methods.

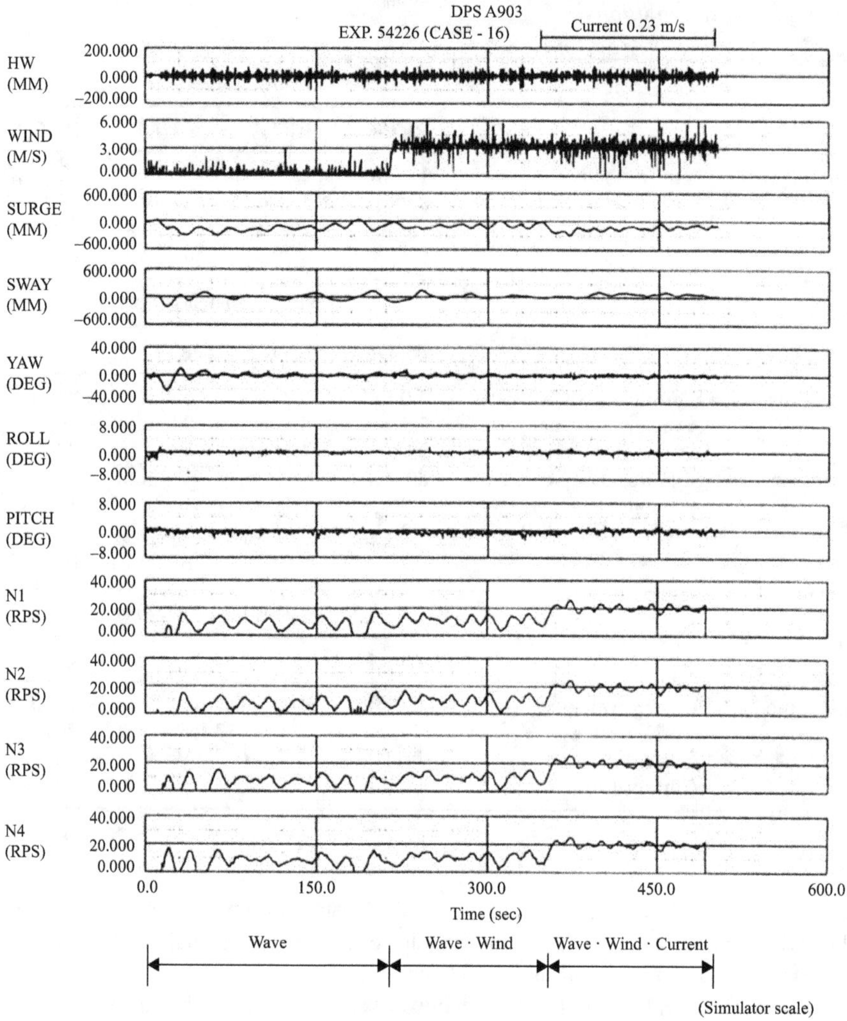

Figure 6.39 DPS test result Beaufort scale 7 (4 thrusters)

This section describes the implementation of the DPS in the Seabex (special model ship) software. The effectiveness of DPS is confirmed by numerical simulation and tank testing using a model ship. This research is an effective approach to designing an offshore platform DPS at low cost. The book is the first one to present this design method.

6.3.6.2 Basic steps of development

Figure 6.44 shows the basic steps, from development of the DPS's basic specifications to execution of the sea trial test. First, the basic specifications, such as the

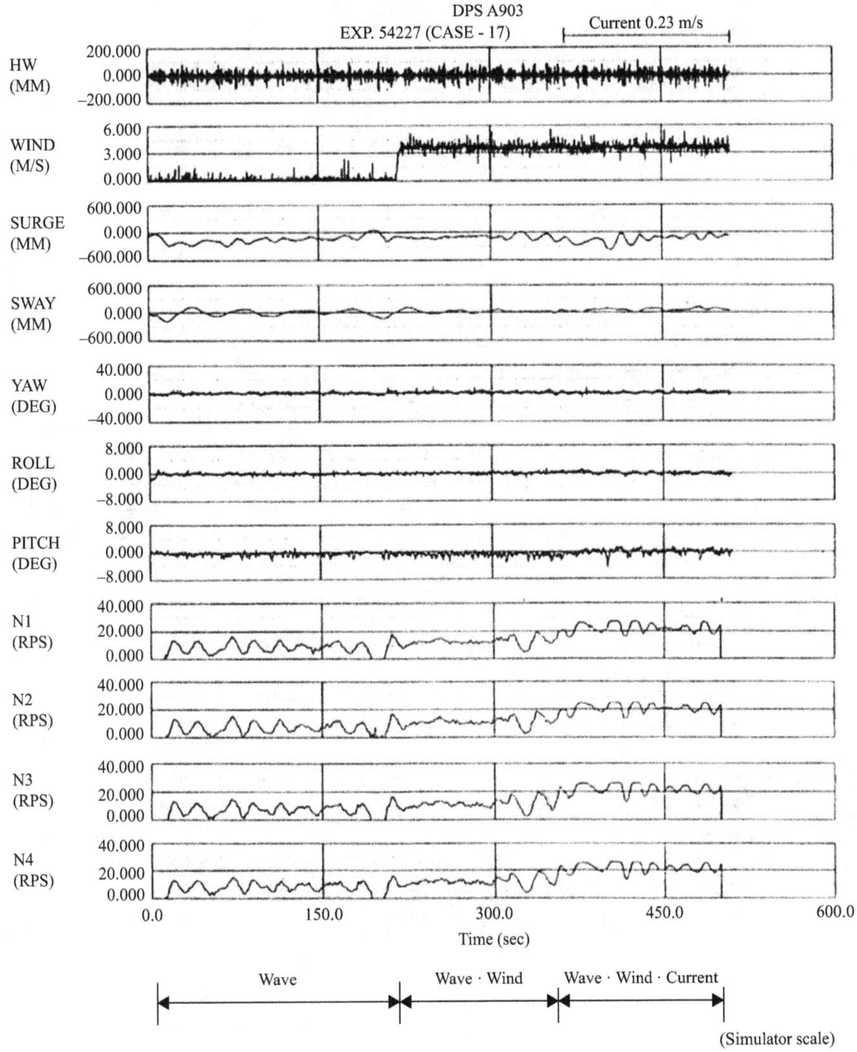

Figure 6.40 DPS test result Beaufort scale 8 (4 thrusters)

system's size, actuators, control system, and sensors, are developed on the basis of the operational requirements. For basic specification items that require the development of new elemental technologies, water tank experiments are conducted using elemental models. For the kinetics system of the model ship, motion simulations are conducted using motion models, and suitable elemental model experiments are planned and conducted. Elemental model experiments are also conducted for items of the instrumentation system that require prior verification, such as control, communication, and navigation. The verification results are then fed back and

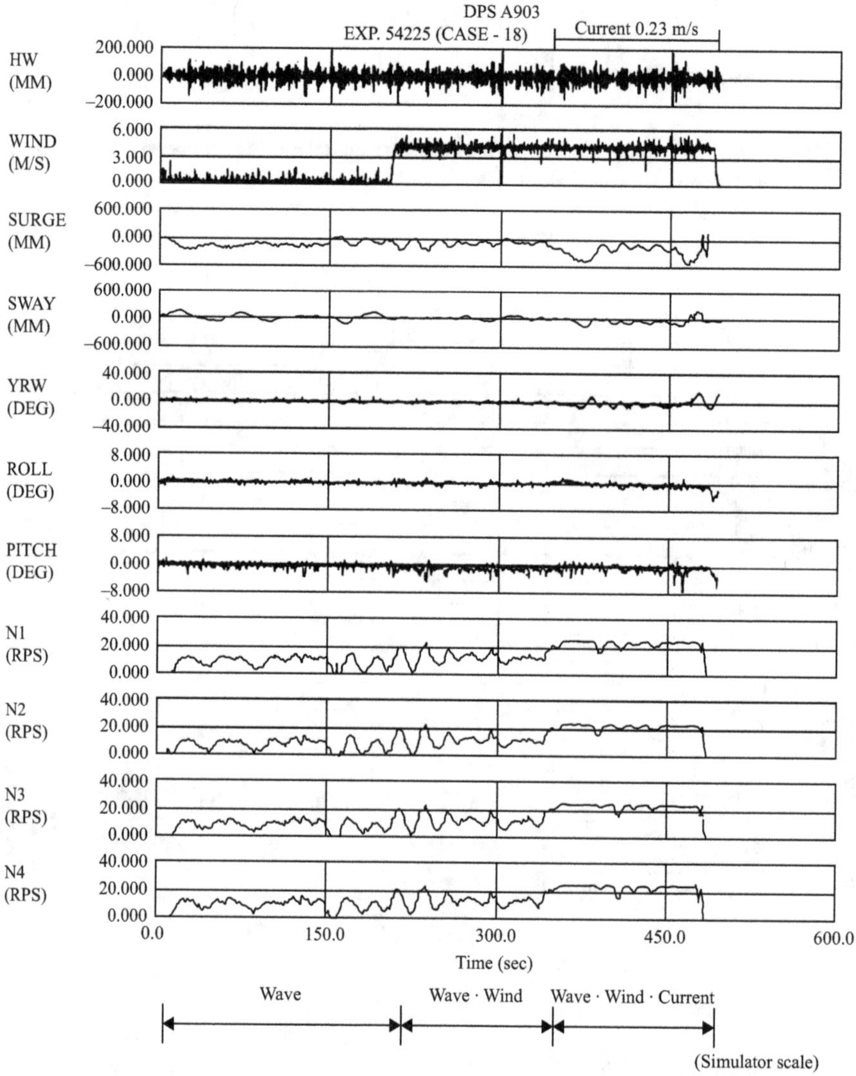

Figure 6.41 DPS test result Beaufort scale 9 (4 thrusters)

reflected into the basic plan. Next, data measurements of the scaled model ship's shape are taken. Techniques for taking data measurements of the shape include modelling methods that use resin materials, and methods that sample stereolithography (STL) data taken with 3D laser measurements. Using this data as computer-aided manufacturing (CAM) data, rapid prototyping is employed to manufacture the model ship. On the basis of the model ship's basic plan, assembly drawings are created through CAD. The assembly drawings are then expanded to create part drawings. The mechanical parts, instrumentation parts, and equipment

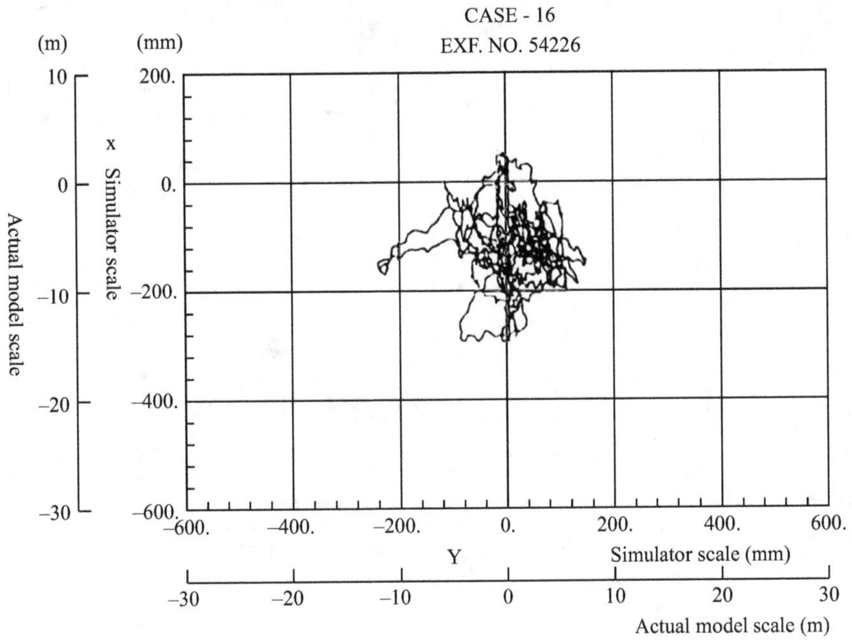

Figure 6.42 DPS test result Beaufort scale 7 (4 thrusters)

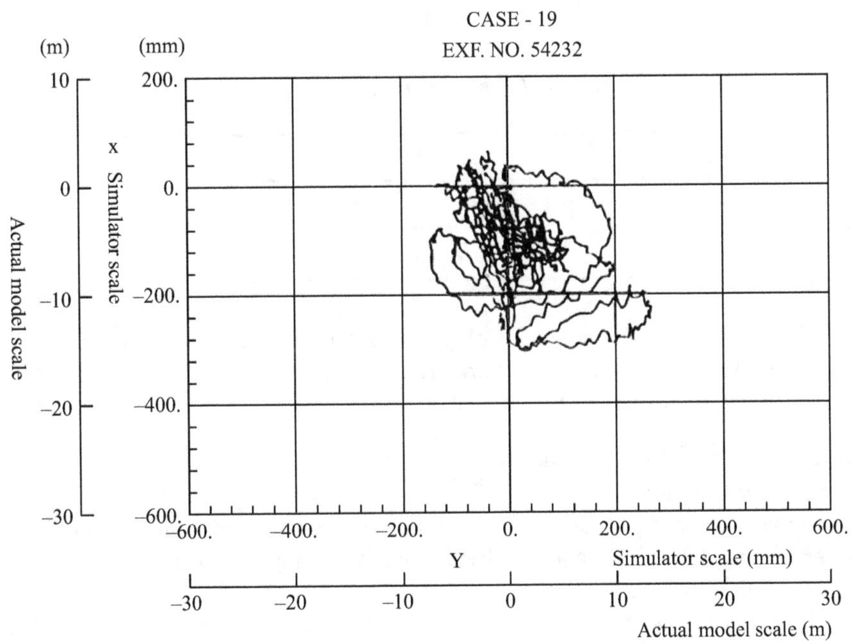

Figure 6.43 DPS test result Beaufort scale 7 (3 thrusters)

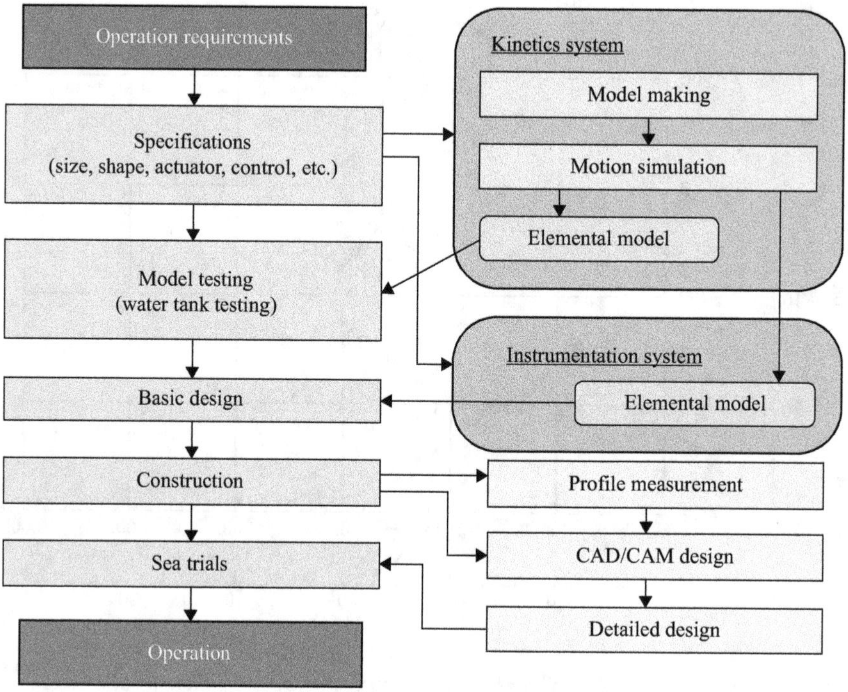

Figure 6.44 Basic steps of development of DPS

Table 6.9 Full-scale characteristics of the model ship

Parameter	**Unit**	
Length	m	106.8
B beam	m	19.6
Draft	m	5.5

are manufactured, and the model ship is assembled. The manufactured model ship then undergoes a sea trial test and, if its functions are confirmed to meet the specifications, development is completed.

6.3.6.3 DPS simulation model

Model ship Seabex

The model ship used is a full-scale model from the Seabex One which was built by the Nobiskrug GMbH shipyard of Rendsburg, Germany. The original parameters are shown in Table 6.9.

To get a first understanding of the DPS model, we start to implement the model ship parameters in the DPS model. Tables 6.10 and 6.11 show the model ship parameters.

Table 6.10 Model test-scale characteristics of the model ship

Parameter	Unit	
Length	m	1.45
B beam	m	0.26
Draft	m	0.078

Table 6.11 Specifications of the thrusters

Parameter	Unit	
Number of thrusters	-	4
Mounting position X-coordinate	m	[−0.32 −0.32 0.575 0.695]
Mounting position Y-coordinate	m	[−0.065 0.065 0 0]
Maximum thruster capacity	kg	[5 5 5 5]

Table 6.12 Optical parameters of the PID controller

X	P gain	0.83
	I gain	0.0065
	D gain	9.53
Y	P gain	0.85
	I gain	0.0067
	D gain	9.78
Z	P gain	0.13
	I gain	0.0011
	D gain	1.56

The model ship is scaled with a length factor of 75. The model speed is scaled with the Froude scaling law.

$$V_{Model} = \frac{V_{Ship}}{\sqrt{\alpha}} = \frac{V_{Ship}}{8.7} \tag{6.19}$$

where α is a scaling factor. The time factor is also scales with Froude scaling law:

$$t_{Model} = \frac{V_t}{8.7} \tag{6.20}$$

All time constants in the model ship must be smaller by a factor of 8.7 than the full-scale ship.

Controller parameters

Basic control calculation was computed by PID control of the decision rule for control force and control moment, as shown in Table 6.11. The author and his colleagues set the PID gains using the PID controller with Simulink®. PID parameters are shown in Table 6.12.

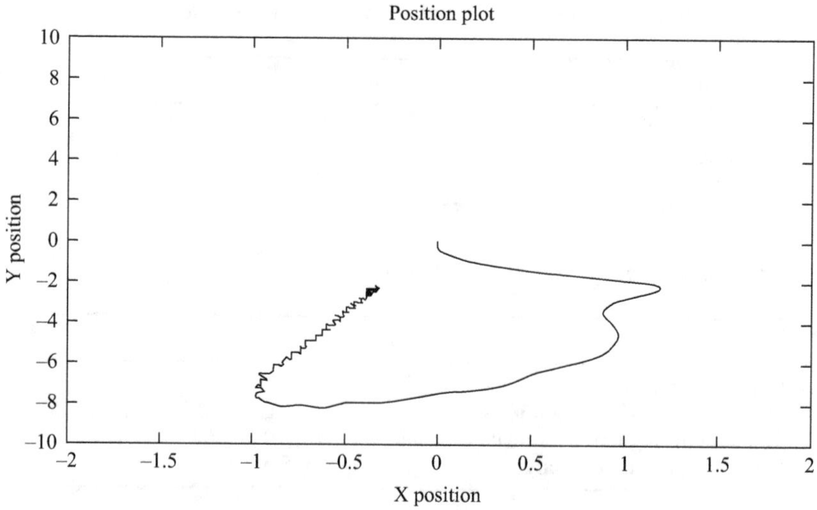

Figure 6.45 Model position by the simulation

Numerical simulations

In the Figure 6.45 the model position is plotted, calculated by simulation. The values are quite large. Probably this has to do with uncertainties of the parameters. Some values are guessed, for example, mass and moment of inertia. The model tank tests show more realistic position plots.

6.3.6.4 Implementation of DPS in the model ship

Figure 6.46 shows a block diagram of the model ship control system. The PID controllers, thruster force allocation, and actuator controllers are a part of the Simulink® software. This software has a USB connection with the Arduino chipset for the controller. The Arduino chipset controls the azimuth thrusters which move the model ship. The position of the model ship is measured with position controllers which are connected via a Bluetooth connection with Simulink®.

Required force control

Figure 6.47 shows all components of the DPS system. The system starts with a position offset. This offset is used as input for the required force controller (see Figure 6.47 block A). This controller consists of three PID controllers, for the X and Y axis forces and the moment around the Z axis. These controllers are exactly the same in the DPS and the model ship software. The operation decision support logic is developed for operation support of emergencies, and linked with the DPS.

Thruster allocation

The thruster allocation of the DPS (see Figure 6.47 block B) is different from the thruster allocation in the model ship software. The thruster allocation consists of

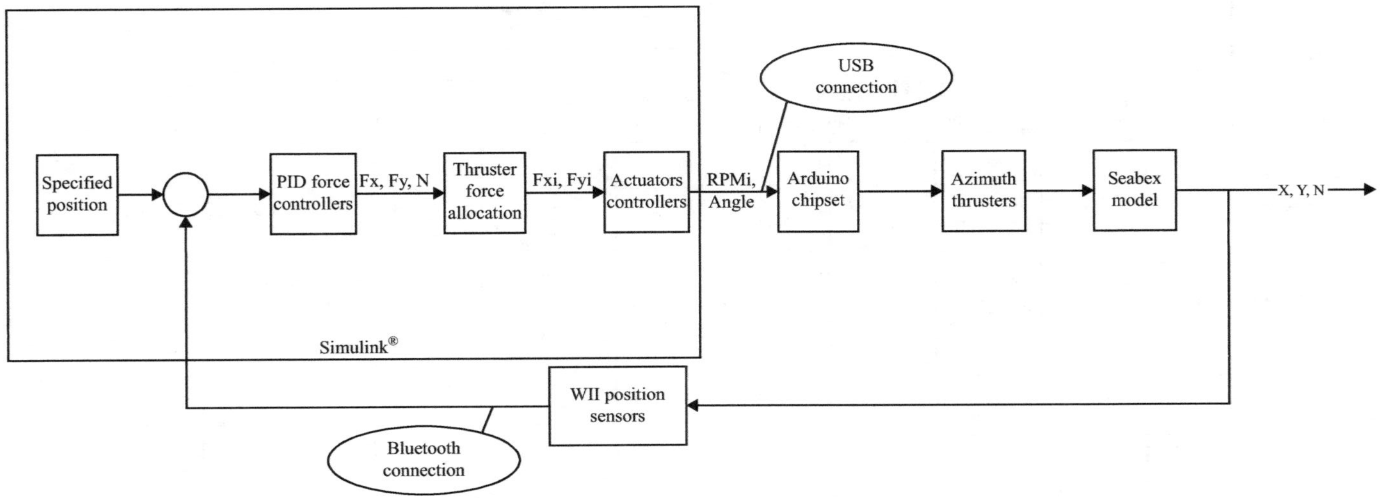

Figure 6.46 Block diagram of model ship control system

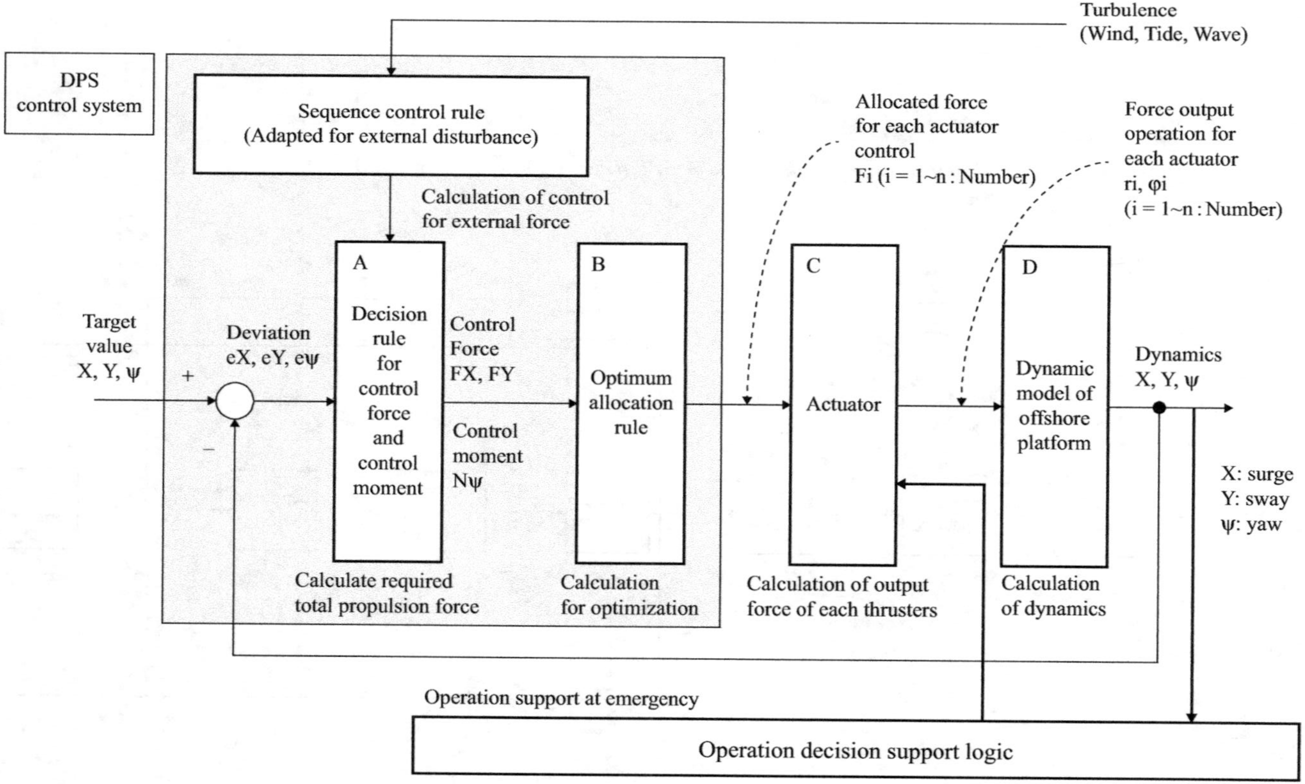

Figure 6.47 DPS block diagram

four MATLAB® function scripts. The first script is for the minimum energy thruster allocation, and the other three scripts are for the optimum angle thruster allocation. Both allocations had 16 output forces in the X direction, forces in the Y direction, moments around the Z axis, and a time delay. This time delay simulated the change of thruster angle and is removed from the model because this delay will be a physical phenomenon in the scaled model ship.

The two allocation functions are implemented in two separate subsystems. Those subsystems can be manually enabled or disabled. This is executed to decrease calculation time.

Actuator

The actuator (see Figure 6.47 block C) is used to transform the required thruster forces in the X and Y directions to a thruster angle and propeller speed. In the DPS simulation model this is simply a sigma function which sums the different thruster forces. However, in the model ship software this part consists of a lookup table to translate a required force to required speed and four speed controllers for the four actuators.

Dynamic model

The dynamic model (see Figure 6.47 block D) covers the main difference between the DPS simulation model and the model ship system. In the DPS model, the dynamic part is simulated on the basis of mathematical equations of movement; the model ship model is based on real physical movements which are returned to the software via position sensors. To test the DPS thruster allocation and fuzzy control in the model ship software environment, the dynamic movement model is implemented in the model ship software.

Fuzzy feedforward control

The DPS simulation model has a fuzzy feedforward controller to improve the ship behaviour with increased disturbances. The fuzzy control is implemented in the model ship software. With the model tank tests, the boundaries of the fuzzy control must be determined.

Model ship GUI

The model ship software is controlled with a general user interface (Figure 6.48). The settings of all controllers can be changed in the 'setting editor'. The thruster allocation can easily change here to check the right allocation. In addition, it is possible to enable or disable the fuzzy control.

6.3.6.5 Tank test

Overview

Tank testing has been conducted in the tank. Figure 6.49 shows an example of a tank test. A laser tracking system above in the figure is set up to determine the position of a model ship. Current and wind generators are equipped with the tank.

Figure 6.50 shows an internal mechanism of a model ship with four thrusters. Bow thrusters in Figure 6.51 and stern thrusters in Figure 6.52 are used for

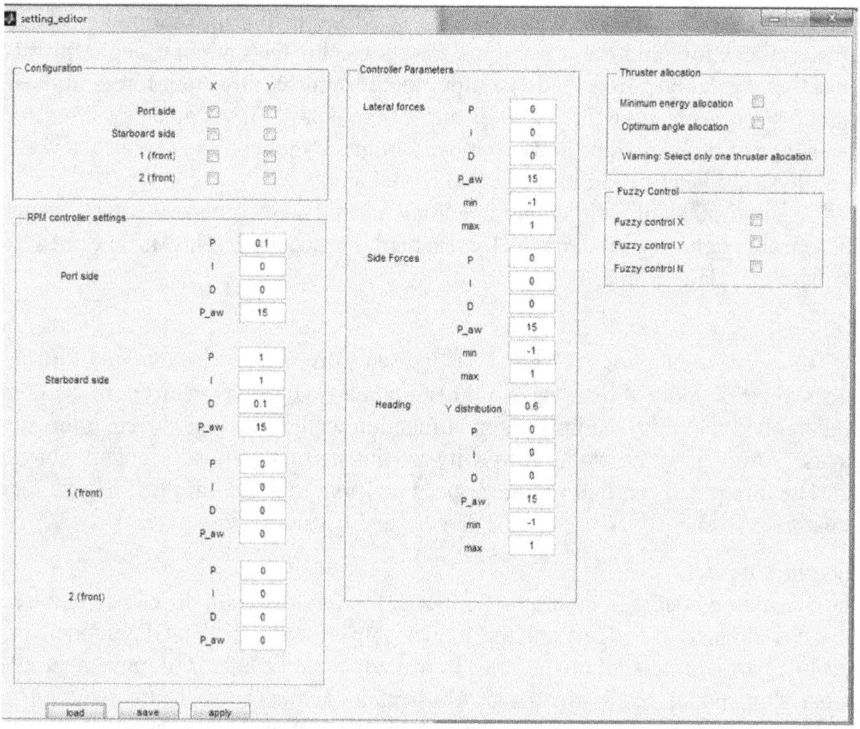

Figure 6.48 Model ship GUI setting editor

actuators. An Arduino chipset (Figure 6.53) is utilized for the controller, and the DPS software is installed in the chipset.

The procedure

The system is tested in test runs of 360 s. The first 60 s are used to stabilize the ship to her position with the different disturbance conditions. The other 300 s of data is logged for the test results. The 300 s on the model scale corresponds to 43.5 min on real scale.

Two different tests were performed; the first tests changed the required position of the model ship to observe the manoeuvrability of the system. The x, y, and heading position where changed every minute in separate runs.

With the other test runs, the position of the model ship was held constant. With these tests, several weather conditions were simulated. Current was simulated with the pump in the fluming tank. The current speed was varied at 0.06, 0.12, and 0.23 m/s, which correspond with a current speed of 0.5, 1, and 2 knots on the real scale. The wind is simulated with three normal table fans, placed to simulate a wide wind field. The fans have a low and high speed setting.

The tests were carried out with all three current speeds, two wind speeds, and all combinations as shown in Figure 6.54. The directions of the wind and current speed were also varied.

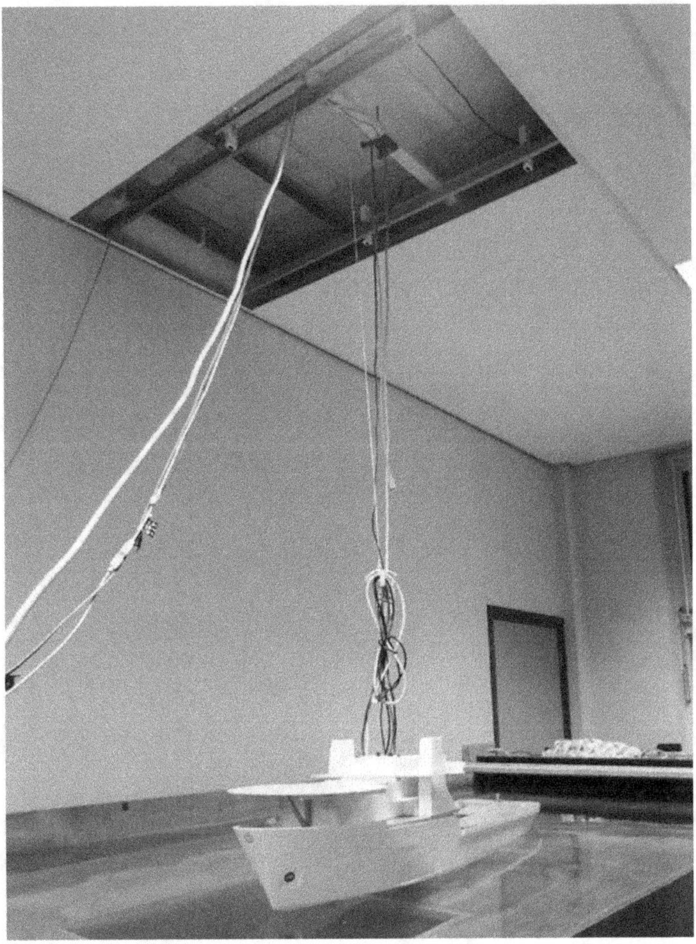

Figure 6.49 Tank test by model ship

6.3.6.6 Results of the tank test

Weather condition tests

Figures 6.55–6.60 illustrate the position-keeping performance under the weather conditions. As shown in Figures 6.55–6.60, the DPS exhibits excellent position-keeping performance.

Cost evaluation

The proposed design method can be used to construct an offshore platform DPS before construction without testing on an actual ship. The author and his colleagues estimate that the offshore platform DPS can be designed at one-tenth of cost of the conventional method.

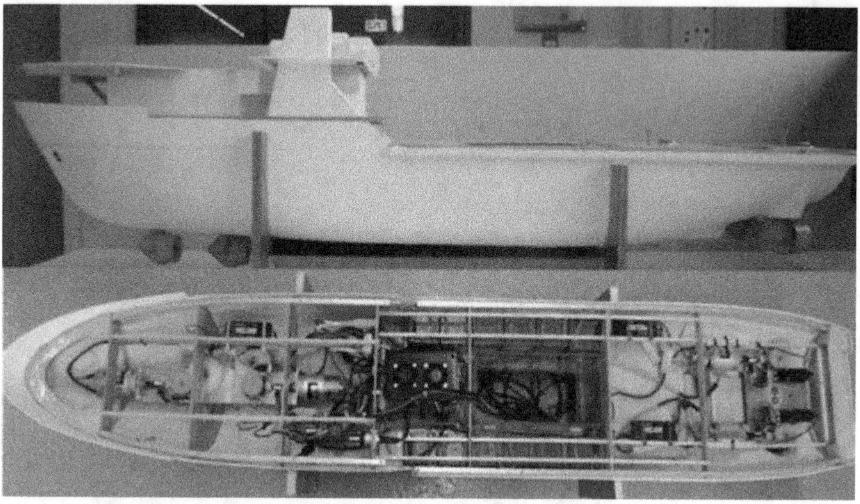

Figure 6.50 Internal mechanism of a model ship

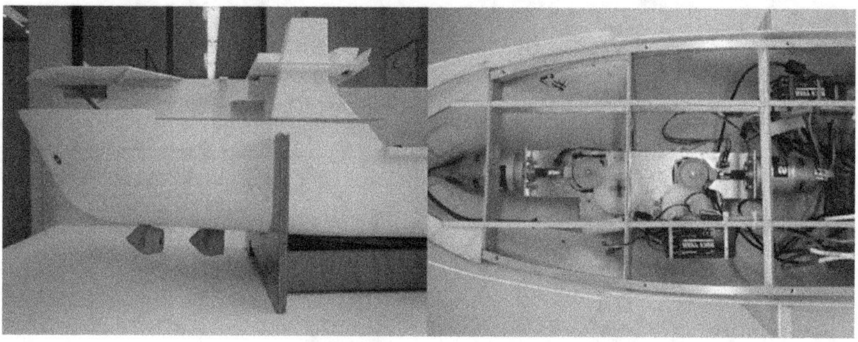

Figure 6.51 Bow thrusters and motors

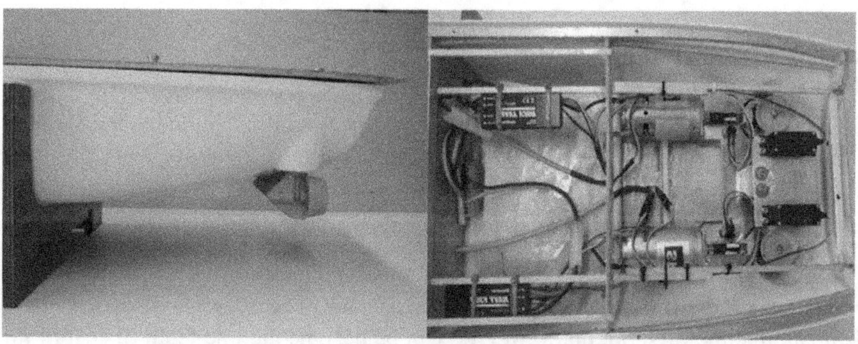

Figure 6.52 Stern thrusters and motors

Figure 6.53 Chipset of the controller

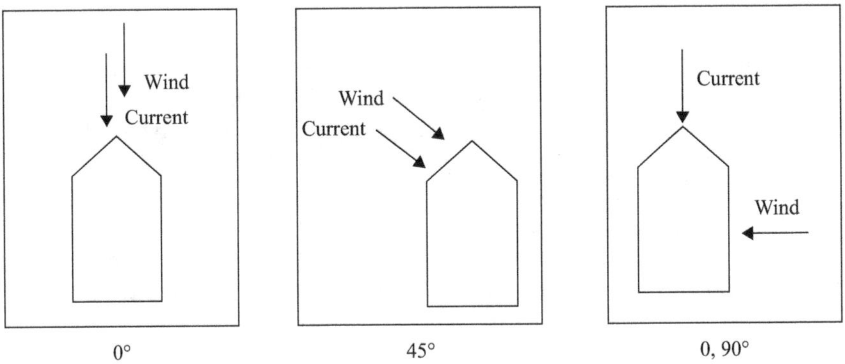

Figure 6.54 Disturbance directions

6.3.6.7 Conclusion

The author and his colleagues developed the DPS for a ship. The system was confirmed by numerical simulation. The effectiveness of the model ship software was confirmed by tank testing. It is also possible that other PID settings must be adjusted for manoeuvring. The author and his colleagues installed the software into a real ship simulator, as shown in Figure 6.61, and will use it in the real ship. The author and his colleagues have established a low-cost design method for offshore

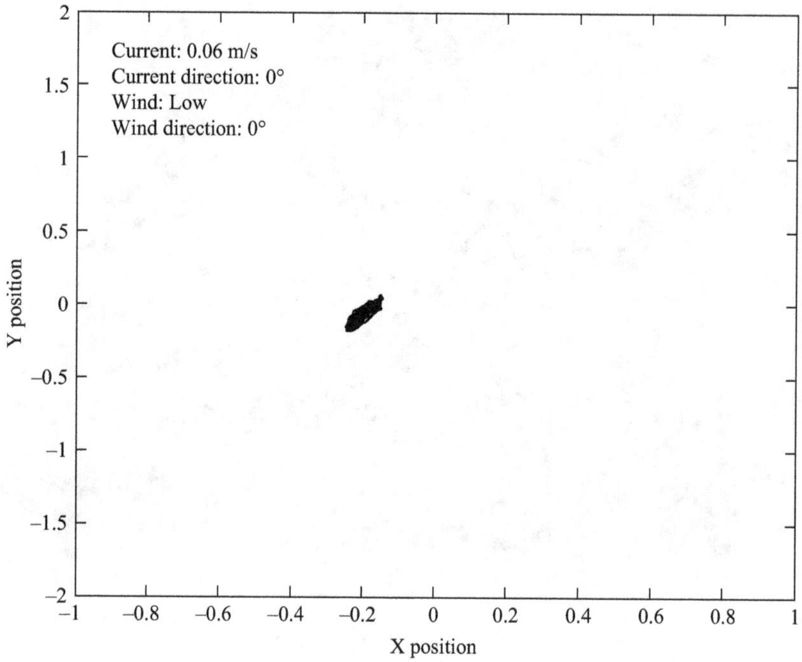

Figure 6.55 Result of tank test (1)

platform DPS before construction at one-tenth of cost of the conventional method. It is possible to perform functionality verification with the use of a low-budget test environment. For instance, it costs more than $1,000,000 to develop the DPS by the conventional design method, but using the proposed design method keeps costs low, at approximately $100,000.

This work was supported by Japan Oil, Gas and Metals National Corporation (JOGMEC). The author and his colleagues would like to express their sincere gratitude to JOGMEC and Delft University of Technology, the Netherlands.

6.4 Robotic fish technology and its new evolution

The author has developed many kinds of robotic fishes, based on an elastic oscillating fin propulsion system, since 1995. This section describes past, present, and future robotic fish technologies, and shows the new evolution of robotic fish applications and technologies in various fields. First, this section summarizes the history of the developed life-like robotic fish, such as sea bream, coelacanth, carp, and shark ray. The robotic fishes are typically propelled by vertical tail fin and operated by servo motors. Second, the robotic dolphin is newly developed. The author and his colleagues developed a tethered robotic whale with a horizontal tail fin propelled by a hydraulic actuator in the 1990s; however, the robotic dolphin is

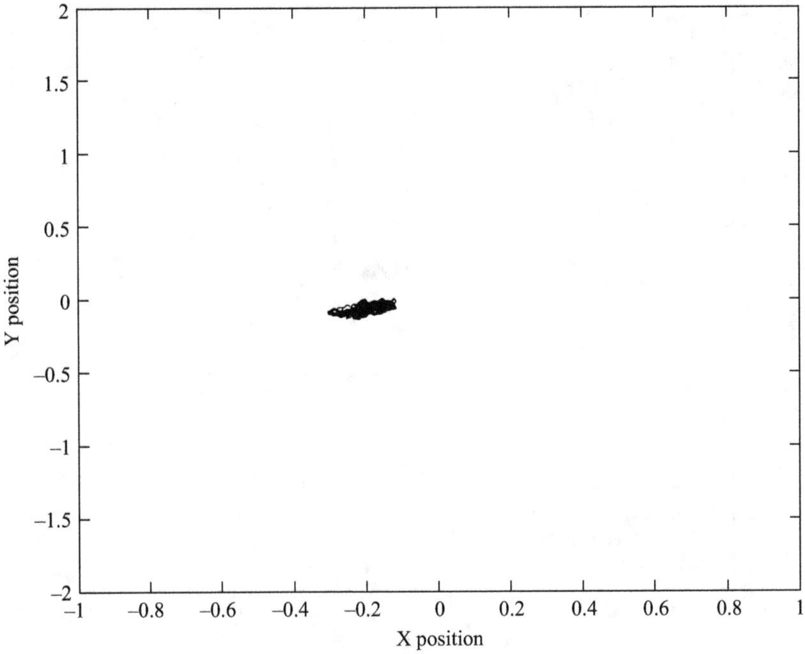

Figure 6.56 Result of tank test (2)

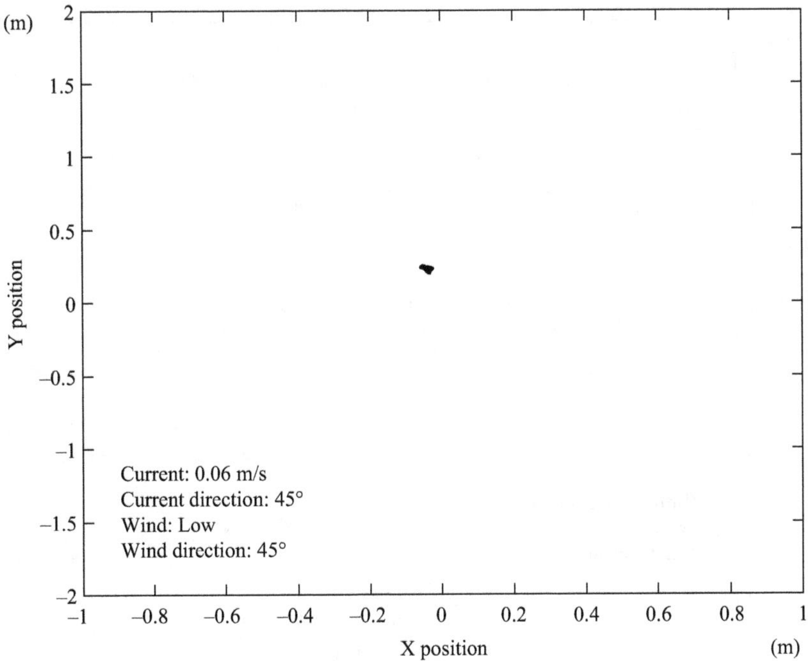

Figure 6.57 Result of tank test (3)

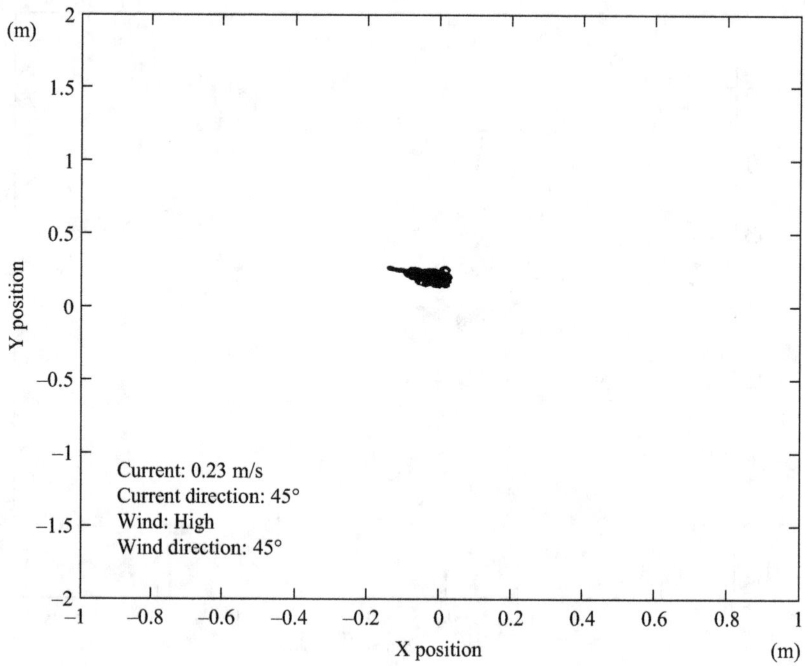

Figure 6.58 Result of tank test (4)

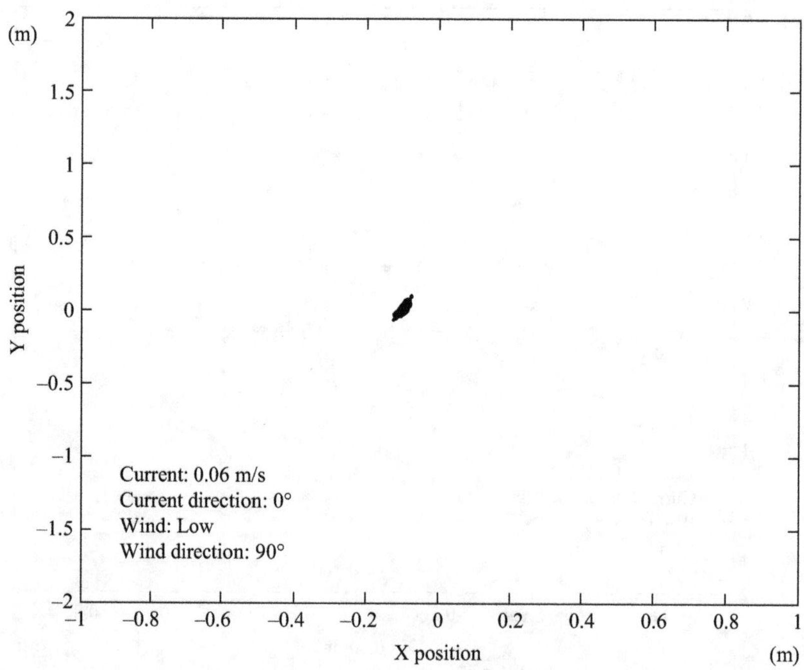

Figure 6.59 Result of tank test (5)

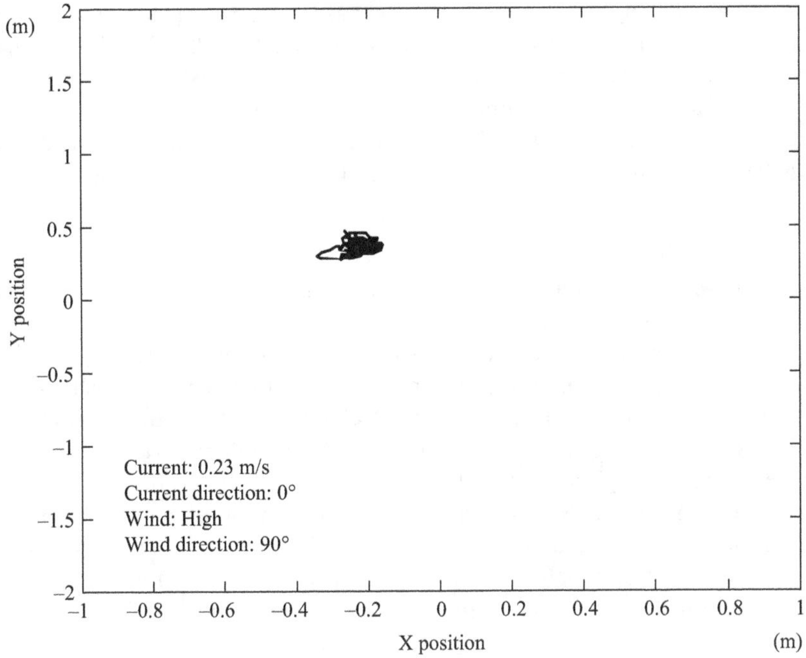

Figure 6.60 Result of tank test (6)

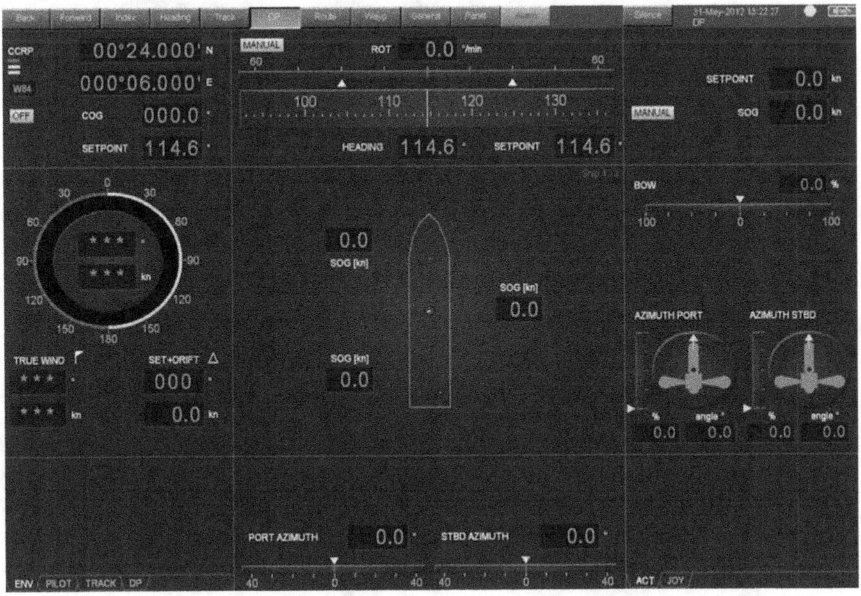

Figure 6.61 Real ship simulator

untethered and is longer (more than 1 m) and has the characteristic of fast cruising and higher manoeuvrability with a horizontal tail fin propelled by servo motors. Third, the book describes new application fields of robotic fish technologies, such as medical forceps and extractors, space robots which went to the International Space Station (ISS) and can work in space, and an ocean current power plant using an oscillating fin propulsion system for new offshore sustainable energy.

6.4.1 Introduction

Creatures have evolved over many generations to adapt to their natural environment. It is possible to create new products and technologies by clarifying their manoeuvring mechanism scientifically and applying this knowledge to engineering.

In particular, an oscillating fin, which is an application of a fish fin to a propulsion device, is the most important technology for creating a new manoeuvring system for marine vehicles. The development of underwater vehicles has been performed through new and different approaches. Above all, observing such marine creatures as fish, and engineering the application of their manoeuvring mechanisms have enhanced the possibility of producing new underwater vehicles. This type of robot is widely known as robotic fish. The first robotic fish were developed from 1993 to 1995 (Figure 6.62) and are widely known. The sea bream

Figure 6.62 First life-like robotic fish with the author in 1995

Figure 6.63 First robotic fish swimming in space

robotic fish is the first developed robotic fish that has gained a good reputation for its life-like swimming without power or communication cables. It is an untethered underwater vehicle. The author and his colleagues have developed over 16 kinds of robotic fish, such as coelacanth, carp, whale, manta, tuna, shark ray, and dolphin. In addition, a space manta swam in space, as shown in Figure 6.63, and applications of the elastic fin's mechanism for grasping organs softly for surgical instruments have been successfully conducted. This section describes the history and updates robotic fish technologies.

6.4.2 History of robotic fish development

The author began to develop an oscillating fin propulsion system as an alternative propulsion actuator to propellers in the 1980s. The flexible oscillating fin propulsion system was created to realize fish-like fin movement. It was found that the flexure like a fish fin was produced by variously changing elastic modules of the flexible oscillating fin. In addition, it was found that the swimming motion of fish varied with living environment, so that the body structure and propulsion method evolve uniquely so as to suit the swimming motion. That is why the author's interest shifted to the creation of robotic fish. The development of robotic fish was promoted through scientific approaches, such as fluid mechanics, materials engineering, vibrational science, tribology, electrical engineering, and control engineering.

A sea bream robotic fish was developed in 1995 as shown in Figure 6.64. The internal structure and system configuration of the sea bream robotic fish are shown in

Figure 6.64 Sea bream robotic fish

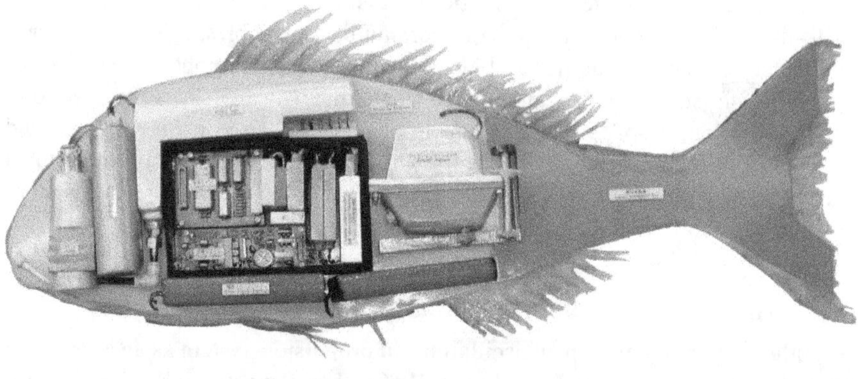

Figure 6.65 Internal structure of the sea bream robotic fish

Figures 6.65 and 6.66, respectively. The weight is 2.6 kg and length is 0.6 m. It has an internal battery and conducts 3D movement without cables. The surface is made of silicone resin. It cruises at 2 km/h autonomously and/or by remote operation. In addition, 1-h cruising was executed in 1995. Figures 6.67 and 6.68 show a swimming test and a flow visualization test, respectively, of the sea bream robotic fish.

A fully automatic swimming system was developed for the purpose of swimming in an amusement facility as shown in Figure 6.69. The robotic fish is located by the ultrasonic sensor on the water tank and automatically controlled by

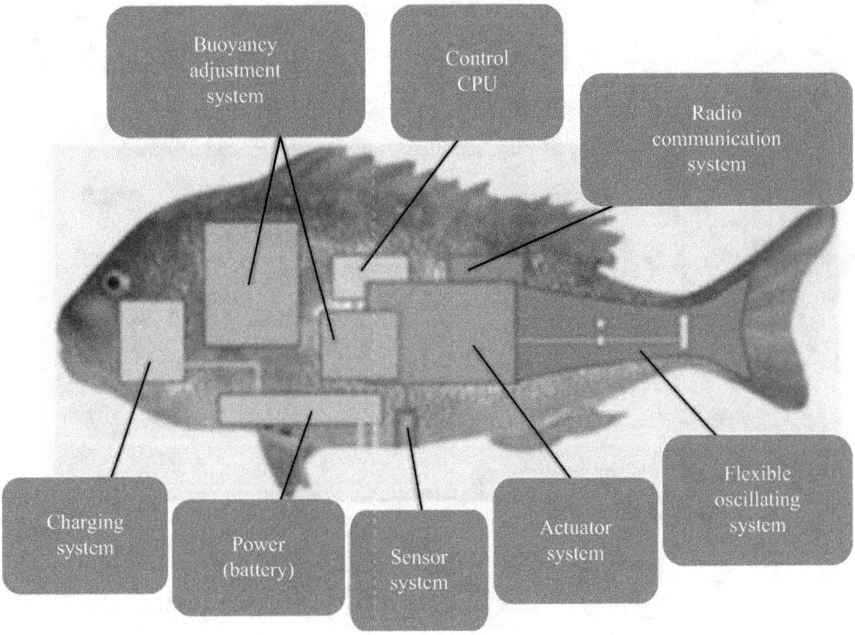

Figure 6.66 System configuration of the sea bream robotic fish

Figure 6.67 Swimming test of the sea bream robotic fish

computer. The control command is sent to the robotic fish by an underwater signal. The system enables the robotic fish to swim continuously. In addition, the author developed an oscillation control method using resonance of fin vibration. The author considers wireless charging for an auto charge system. The robotic fish

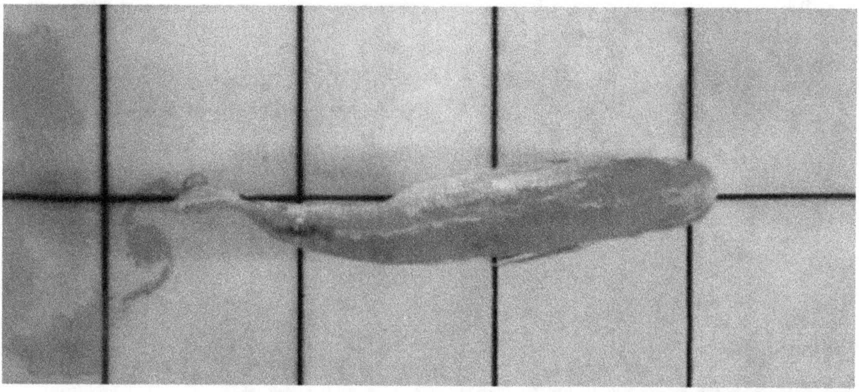

Figure 6.68 Flow visualization test of the sea bream robotic fish

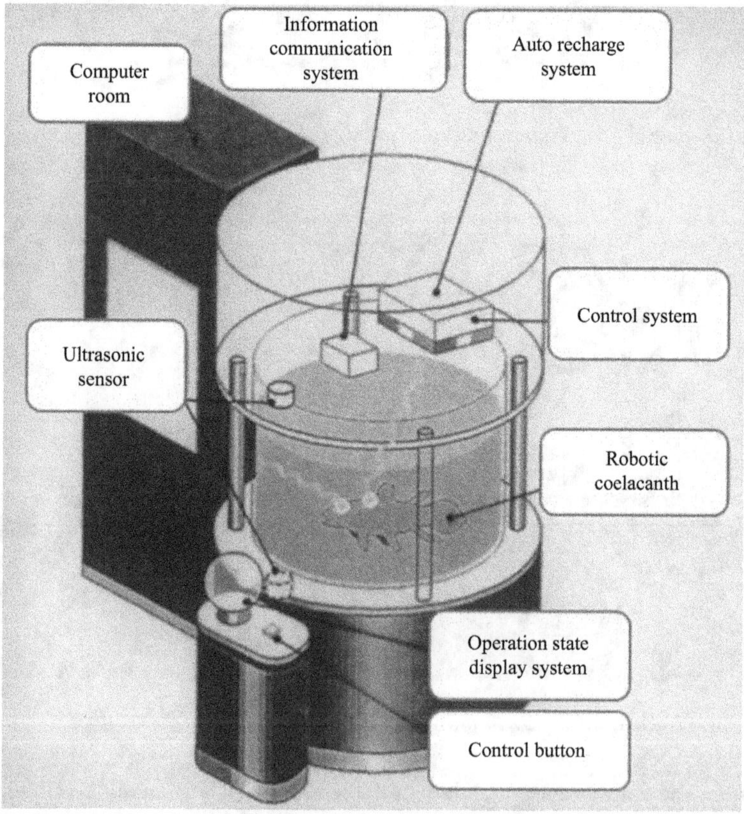

Figure 6.69 Automatic swimming system of a robotic fish

Figure 6.70 Coelacanth robotic fish

starts swimming automatically when the control button is pressed, automatically controls its depth and course, swims to a recharger when its battery runs low, and starts swimming again after automatic charging, as shown in Figure 6.69. The coelacanth robotic fish swims in the water tank of a museum, as shown in Figure 6.70. As the robotic fish has a central processing unit (CPU) in its body, it can swim autonomously.

Additionally, basic steps of development can be established as shown in Figure 6.71. First, the specifications, such as the robotic fish shape, size, elasticity, and material characteristics, are designed on the basis of its operational requirements. Development of new elemental technologies is required to determine basic specification items. For the kinetics system of the robotic fish, motion simulations are conducted using motion models in CAD, and suitable elemental model experiments are also conducted for items of the robotic fish that require prior verification. Next, verification results of the data measurements of robotic fish are taken. The process is considered important in reproducing an effective appearance or in developing a robotic fish whose appearance is convenient for use. Techniques for taking data measurements of the outer shape include modelling methods that use materials, and methods that use STL data taken with 3D laser measurements. Using the data as CAM data for a 3D printer, rapid prototyping is employed to manufacture the outer shape of the robotic fish. On the basis of the basic plan, assembly drawings are created through CAD. The assembly drawings are then expanded to create part drawings. The mechanical parts, instrumentation parts, and equipment parts are manufactured, and the robotic fish is assembled. The robotic fish then undergoes swimming tests, and, if its functions are confirmed to meet the specifications, development is completed.

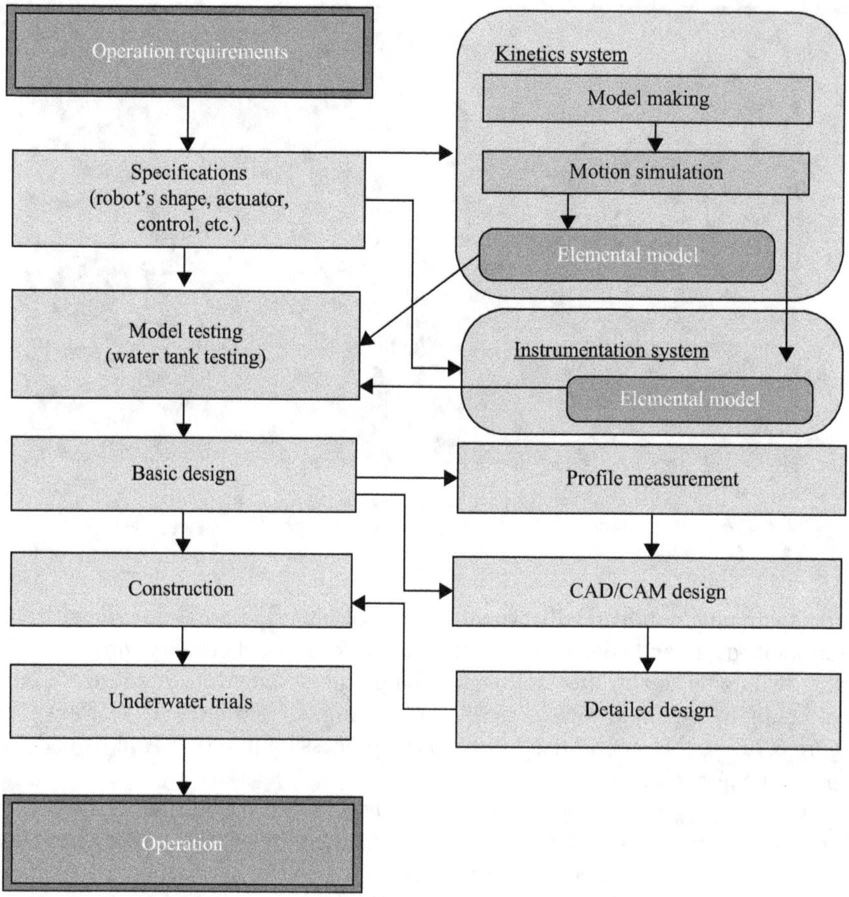

Figure 6.71 Basic steps of development for robotic fish

The author and his colleagues developed various types of robotic fish as shown in Figures 6.72–6.74, for example, on the basis of these steps and the developed robotic fish technologies.

6.4.3 New evolution of robotic fish

The robotic fish has made a new evolution by its uniqueness recently. One is shark ray robotic fish, the other is robotic dolphin.

The robotic fish which has wings and a tail fin has been developed as shown in Figure 6.75. It is called a shark ray robotic fish. The wings of a shark ray have the function of flapping, and are effective for turning stability; the tail fin has the function of propulsion and rudder, as shown in Figure 6.76. The robotic fish has higher manoeuvrability, such as turning at a point. Figure 6.77 shows an external view of a prototype underwater vehicle with a single plate fin mechanism used on performance evaluation tests. Two backward motors for the tail fin provide

Figure 6.72 Carp robotic fish used in Japanese TV as a main actor called 'Shogun'

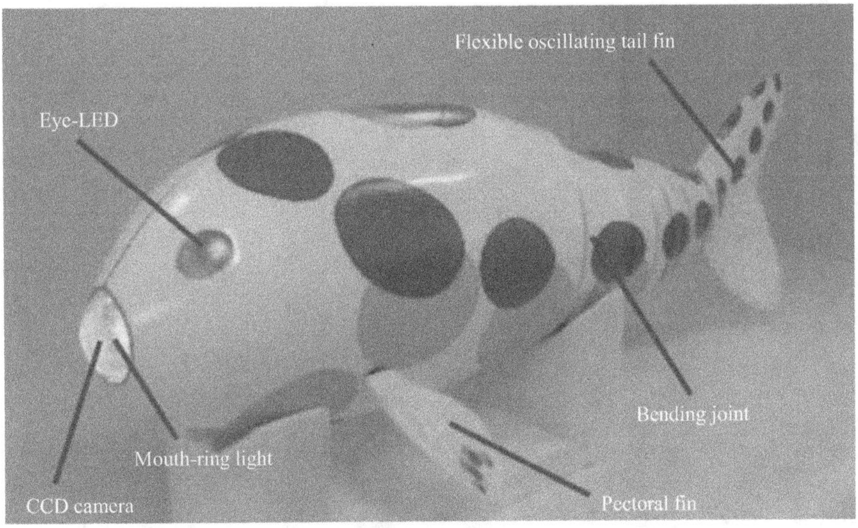

Figure 6.73 Carp robotic fish

propulsion thrust force and two forward motors for the flapping fin maintain the balance of the robot as a stabilizer.

The other evolution is to make a robotic fish with a large vertical tail fin operated by motors. The author and his colleagues have developed the robotic dolphin shown in Figure 6.78. The length is 1.03 m, and the propulsion is produced

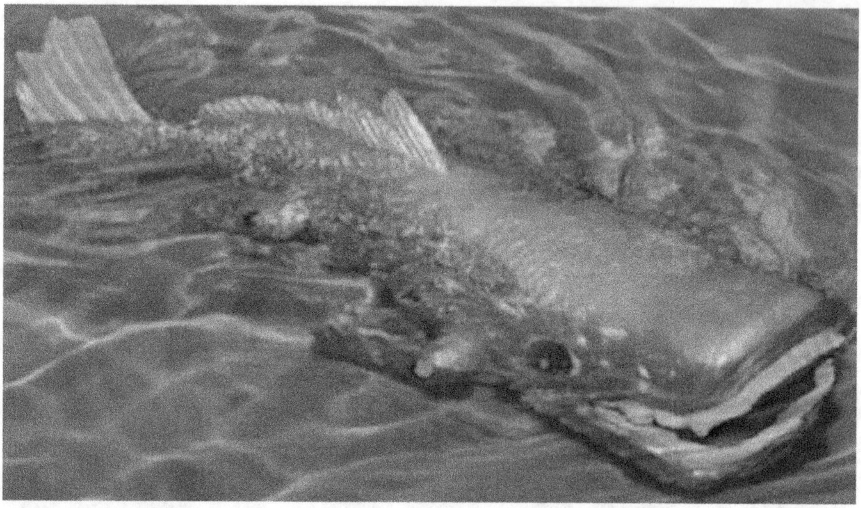

Figure 6.74 Golden robotic Kinschachi for Nagoya EXPO, Japan in 2006

Figure 6.75 Shark ray robotic fish

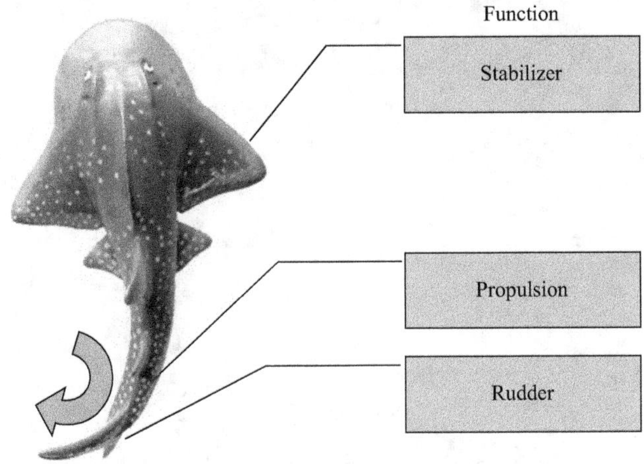

Figure 6.76 Structure of the shark ray

Figure 6.77 Internal structure of the shark ray robotic fish

by motors. In the robotic dolphin, eight servo motors are used for the tail fin and two servo motors are used for the breast fin. The author and his colleagues also developed a robotic whale in 1998, with a length of 4.50 m and powered by hydraulics. The robotic dolphin is the first large robotic fish operated by a vertical tail fin and motors. The cruising speed is approximately 10 km/h.

6.4.4 Advances in robotic fish technology for space, medicine, and offshore energy systems

The robotic fish technology is applied to flexible-structure robotics in space. The robotic fish swam in space and its certification of cruising by NASA is shown in Figure 6.79. It was the first event in space that shows that an elastic oscillating

Figure 6.78 Robotic dolphin

robot is effective for manoeuvring in space. The flexible structure is not only for effectiveness of high manoeuvrability and safety against collisions, but also for a soft-grasping manipulator in space.

Another effective application field of robotic fish technology is medical engineering. There has been an increasing need of surgical instruments that can grasp organs softly and stably. The instrument brings effective surgical operation by decreasing the physical and mental strain of both surgeons and patients. The fish mechanism produces a new soft-handling forceps, as shown in Figure 6.80. The forceps can reduce by 60% the force acting on organs compared with a conventional forceps. In addition, the fish skin mechanism which has the effect of reducing hydrodynamic resistance is effective for creating surgical instruments that are not made slippery by blood. Figure 6.81 shows an example of the new forceps shaped by shark skin. Thus, surgeons and patients can have flexible grasping tools such as an extractor and forceps. The flexible grasping surgical instrument and nonslippery surgical instrument based on robotic fish technology are effective in reducing the workload of treatment, which is important in surgical operations.

In addition, an application of the oscillating fin mechanism is a tidal force generator for an offshore energy system, as shown in Figure 6.82.

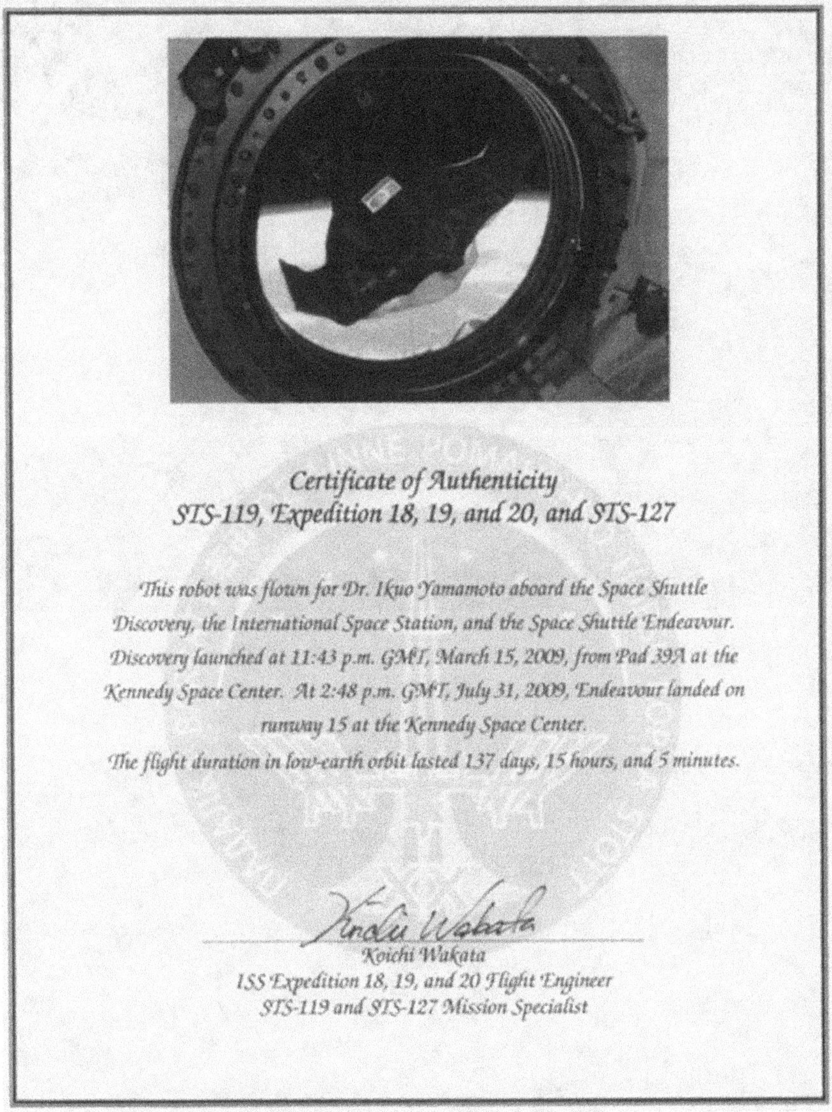

Certificate of Authenticity
STS-119, Expedition 18, 19, and 20, and STS-127

This robot was flown for Dr. Ikuo Yamamoto aboard the Space Shuttle Discovery, the International Space Station, and the Space Shuttle Endeavour. Discovery launched at 11:43 p.m. GMT, March 15, 2009, from Pad 39A at the Kennedy Space Center. At 2:48 p.m. GMT, July 31, 2009, Endeavour landed on runway 15 at the Kennedy Space Center.
The flight duration in low-earth orbit lasted 137 days, 15 hours, and 5 minutes.

Koichi Wakata
ISS Expedition 18, 19, and 20 Flight Engineer
STS-119 and STS-127 Mission Specialist

Figure 6.79 Certification of robotic space manta swimming in space

6.4.5 Conclusion

The author first developed the elastic oscillating fin propulsion system of robotic fish, and over 16 kinds of robotic fishes have been created by the author. The effectiveness of the elastic oscillating fin propulsion system is certified through these developments. In addition, robotic fish technology is applied to new fields,

Figure 6.80 Soft grasping forceps created by fin mechanism

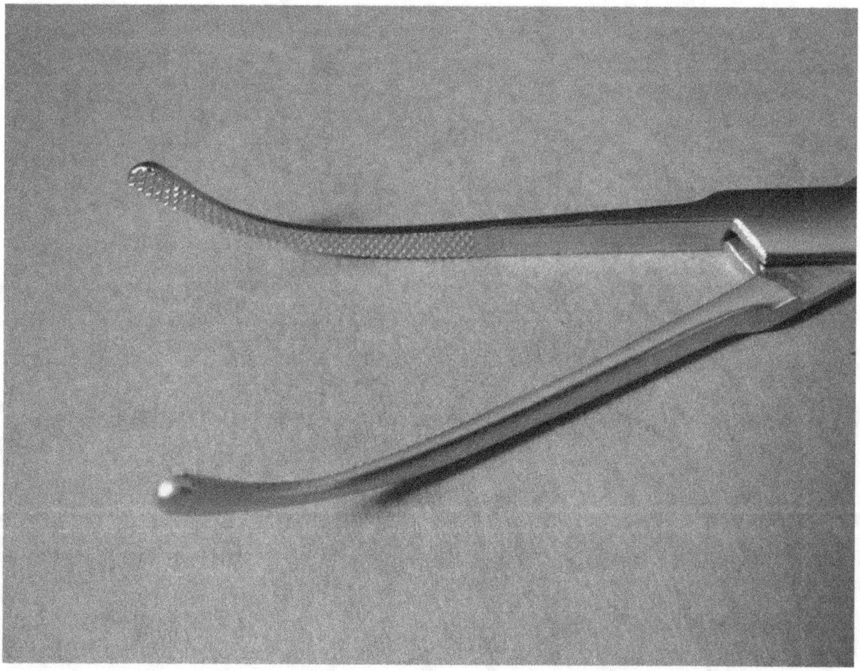

Figure 6.81 Nonslippery forceps by shark skin mechanism

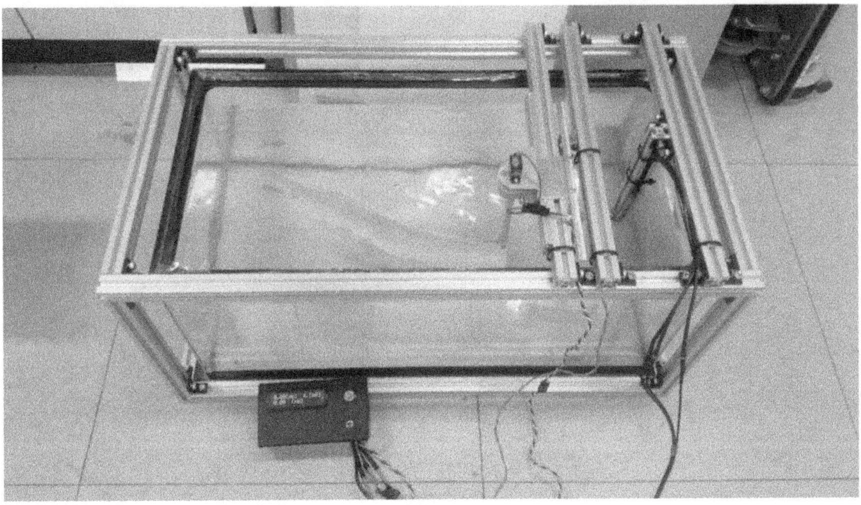

Figure 6.82 Tidal force generator for an offshore energy system

such as space, medicine, and offshore energy systems. Fruitful creation of new robotics and instruments are also expected in the future.

The author plans to make a new type of robotic fish which is effective for fishery and biological investigation research, and a robotic mermaid which has working hands and propulsive tail fin.

Acknowledgement

The author would like to express his sincere gratitude to the Ministry of Education, Culture, Sports, Science and Technology, Japan, JAMSTEC, Mitsubishi Heavy Industries, Ltd., and Nagasaki University for supporting the research.

Chapter 7

Advances in aerospace robotics and mechatronics

7.1 Aviation robotics

7.1.1 Introduction

This research concerns the construction of a multifunctional observation and forwarding system for disaster situations, as a prompt information transmission system when a disaster occurs. Primarily, the development of the observation and forwarding system which can transfer image and sensor information using the Internet is introduced. This means no expensive special equipment is required. Moreover, it can offer a variety of observation forwarding systems that match the needs of a disaster area, which change hour by hour, by combining small flight systems which have separate special functions. When disasters such as earthquakes, typhoons, and tsunamis happen, quick responses to maintain vital services are required, and electricity is especially important. In the conventional way, when a utility pole in a mountain or island area has been damaged, for example, the disaster area should be surveyed to check damage conditions using a manned helicopter or similar means; however, it might be difficult to operate a helicopter at a sufficiently low altitude because of severe weather conditions such as strong wind. Thus, instead of a manned helicopter, a safer, more accurate, faster, and lower cost method to obtain information regarding the damaged area is required. The author and his colleagues developed a low-cost unmanned helicopter and real-time video data transmitting system. We also conducted field experiments on this system. Because the proposed system can perform unmanned observation, there is no risk for causing a second disaster. Even if the system should become damaged, because of its small size and a low cost, the observation risk can be reduced to a minimum.

7.1.2 Construction of the system

7.1.2.1 Summary of the system

A radio-controlled helicopter was selected as an unmanned flying machine for the new system. Although devices such as radio-controlled airplanes, balloons, and kites were examined, the radio-controlled helicopter, which can keep stable flight view with low cost, was found to be the best machine to obtain correct information of the disaster area. The radio-controlled helicopter is equipped with partly

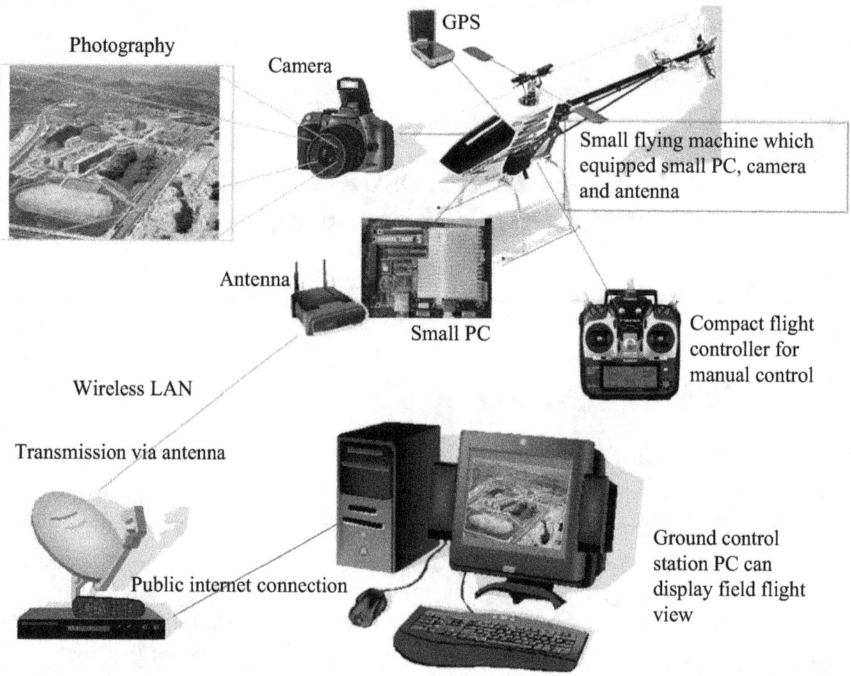

Figure 7.1 Summary of the system

automatic control (feedback control based on displacement of the yaw angle which a gyroscope unit detects) of the rotation angle. This helicopter is controlled by an expert user, and autonomous control by GPS is not installed. The author and his colleagues also developed a real-time video data transmitting system through a wireless LAN and incorporated it on the radio-controlled helicopter. A summary of the system is shown in Figure 7.1.

7.1.2.2 Design of video data transmitting system

An illustration of the system design for the video data transmitting system is shown in Figure 7.2. The video camera generates an analogue video signal while recording HD video. Next, this system encodes the analogue video signal into a digital internet video data and transmits it across a wireless local-area network (LAN). Here, the author and his colleagues equipped a small-board PC in a design for the expansion of observations such as temperature, humidity, altitude, and run trace. This PC was also connected to the wireless LAN, and all of the data from the PC can be received with the video data at the base station. This system can show real-time video images of a disaster area to the ground base station and transmitted to a home PC monitor via a public Internet connection. When the disaster location is far away from the base station, the system communicates using a long distance specification access point and an added loop antenna for the base station.

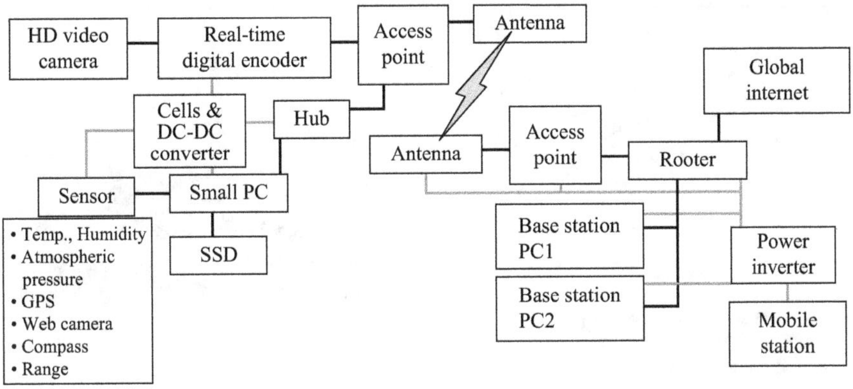

Figure 7.2 System design of the data transmission device

Table 7.1 Specification of the device

Full length	1570 mm
Overall height	670 mm
Weight	6600 g
A payload	3000 g
Flight hours	30 min.
The gear ratio	8:1:5.18
Power source	2st. gasoline engine
Displacement volume	30 cc
A tail drive type	Belt drive
A mainframe	Duralumin Upper: 2 mm, Lower: 3 mm

7.1.3 Production of the hardware

7.1.3.1 Main body of the helicopter

The device specification is shown in Table 7.1. There is a payload limit of approximately 3 kg in addition to the weight of the helicopter main body. It is necessary to produce and install a video transmitting system within the 3-kg payload limit.

7.1.3.2 Production of the video data transmitting unit

This unit consists of a video data encoding device, small PC to observe (for extending), battery, transmitting device (wireless LAN), and two antennas. Because of the payload limit of the main body of the copter, video data transmitting unit was produced within 3 kg. The battery, PC, and the video encoder are arranged for right-to-left balance. The case was produced from 0.3-mm thickness aluminium board to reduce the weight. The weight in this part is approximately 2 kg. This unit is shown in Figure 7.3.

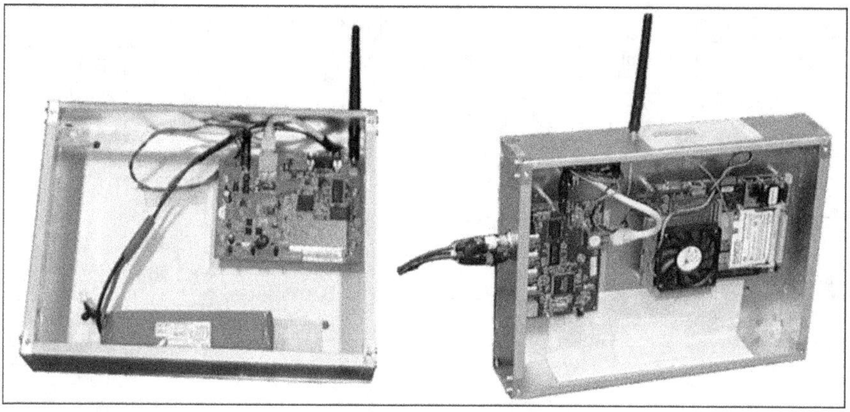

Figure 7.3 Video data transmitting unit

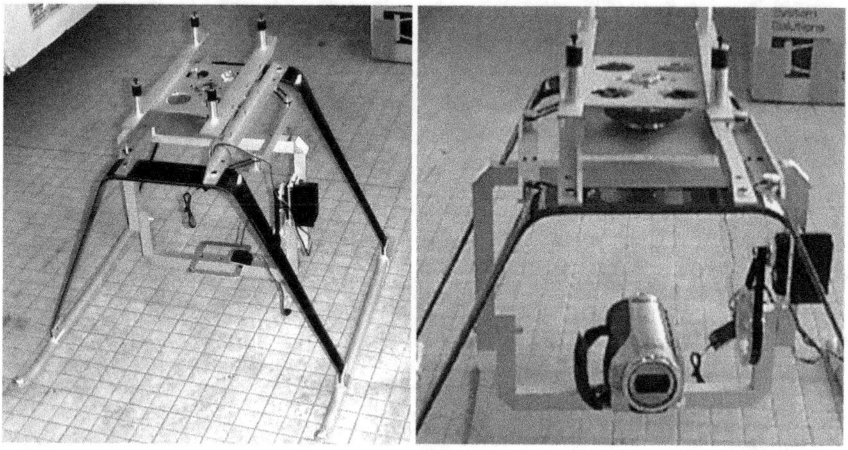

Figure 7.4 Skid

7.1.3.3 Equipping the camera

The HD video camera is installed for Full HD (1920 × 1080 pixels) recording and analogue video signal (640 × 480 pixels) output. The camera is fixed to the skid (camera movable plinth). Figure 7.4 shows the skid.

7.1.3.4 Equipping the video data transmitting unit

In the final assembly, the video data transmitting unit was equipped beside the skid. The helicopter assembled as an observation robot is shown in Figure 7.5.

7.1.3.5 Installation of the base station

The base station is set up in a car for mobile operation. The power supply is 100 VAC converted from the 12-VDC car battery. Figure 7.6 shows the base station.

Figure 7.5 Flying observation robot

Figure 7.6 Base station

7.1.4 Operation experiment

7.1.4.1 Operation experiment A

The operation experiment of this system was performed at an airport (Soeda-cho, Fukuoka Prefecture, Japan) for the model aircraft in March, 2008. Figure 7.7 shows the appearance of the experiment. The experiment succeeded in real-time transmitting of video images through the wireless Internet standard. The transmitting distance

Figure 7.7 Appearance of the experiment

between the helicopter and the base station was approximately 100 m. The quality of transmitted video image was excellent (640 × 480 pixels, 30 fps, approximately 0.5-s delay).

7.1.4.2 Exploratory experiment

The author and his colleagues experimented to investigate the limit of transmitting distance before operation experiment B. Figure 7.8 shows the result. In the experiment, the video data transmitting unit was detached from the body of the helicopter. The distance limit to maintain high-quality real-time video image was 1.5 km (line-of-sight).

7.1.4.3 Operation experiment B

The author and his colleagues participated in an experiment for a surveillance study for the Kyushu Bureau of Telecommunications, Ministry of Internal Affairs and Communications, in October, 2008. The experimental location was Joyo-cho, Yame City, Fukuoka Prefecture, Japan. This region is a mountainous district where the influence of natural damage (such as roads cut by earth and sand, and downed trees) frequently occurs, and is an inconvenient region far from any urban area. The experiment confirmed that the system could transmit video data via Internet in a simulated disaster area. Figure 7.9 shows the appearance of this experiment. Figure 7.10 shows a picture from HD video data recorded on the flying helicopter. Figure 7.10 shows that it was possible to take a picture of cooking smoke from the sky. Therefore, it seems that a forest fire could surely be found.

Figure 7.8 Measurement of distance transmitting video data

7.1.5 Disc-type flying robot

The author and his colleagues want to equip a flying machine with an infrared camera to operate at night. Moreover, the author and his colleagues have developed a new type flying robot since 2007, because the current robot is difficult to operate and manoeuvre to a controlled position as shown in Figures 7.9 and 7.10.

It is a disc-type flying robot and has eight brushless motors (Figures 7.11 and 7.12). Its weight (when equipped with the camera) is approximately 2 kg and its maximum payload is also approximately 2 kg. Its flight time in 2008 was approximately 15 min. As the perimeter of rotors is a disk type (original structure), a chance of collision with an object does not directly impact the rotor blades. Compared with the helicopter used by the operation experiment, this machine is small and cheap with stable flight. Its shape was developed to enhance flight stability on the basis of experiments in which the current of air generated when flying was made visible.

7.1.6 Compact flying boat robot

The author and his colleagues altered a compact flying boat robot to mount observation equipment (such as a camera and various sensors), and conducted

Figure 7.9 Appearance of the experiment to transmit video data

Figure 7.10 Appearance of the experiment to take a picture of smoke

Figure 7.11 Disk-type flying robot in Tokyo

Figure 7.12 Appearance of the experiment of the disc-type flying robot to observe electric cables

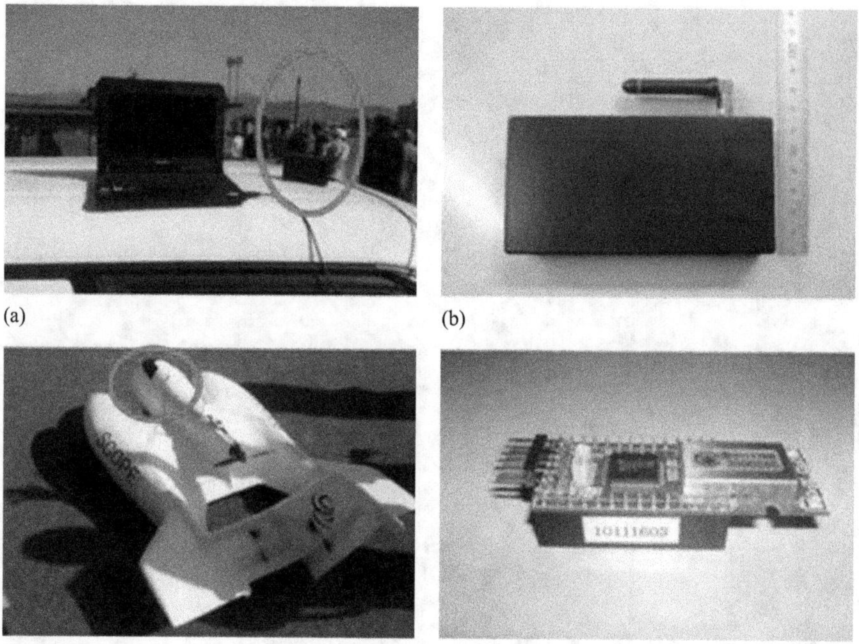

(a) (b) (c) (d)

Figure 7.13 Wireless transfer unit: (a) base unit, (b) handset unit, (c) antenna of the handset unit, (d) wireless module (950 MHz)

experiments. A specified low-power wireless transfer unit was developed for mounting on the robot. This wireless transfer unit operates in the 950-MHz band. Figure 7.13 shows the wireless transfer unit.

The author and his colleagues conducted verification experiments for the compact flying boat robot which includes the new wireless transfer unit. Figure 7.14 shows the verification experiments.

7.1.6.1 Manoeuvrability and stability

The compact flying boat robot including the new wireless transfer unit took off from a tight spot, and landed on a tight spot. In addition, the robot stably made flights and transmitted video pictures to the ground in real time. The optimization of the direction of the camera is an issue for future development.

7.1.6.2 Capability and packet arrival rate

The author and his colleagues measured the packet arrival rate by measurement software which was developed by the author and his colleagues. Table 7.2 shows the results of measurement. An implementation of re-forwarding communication protocol and an efficiency of the antenna are issues for future investigation.

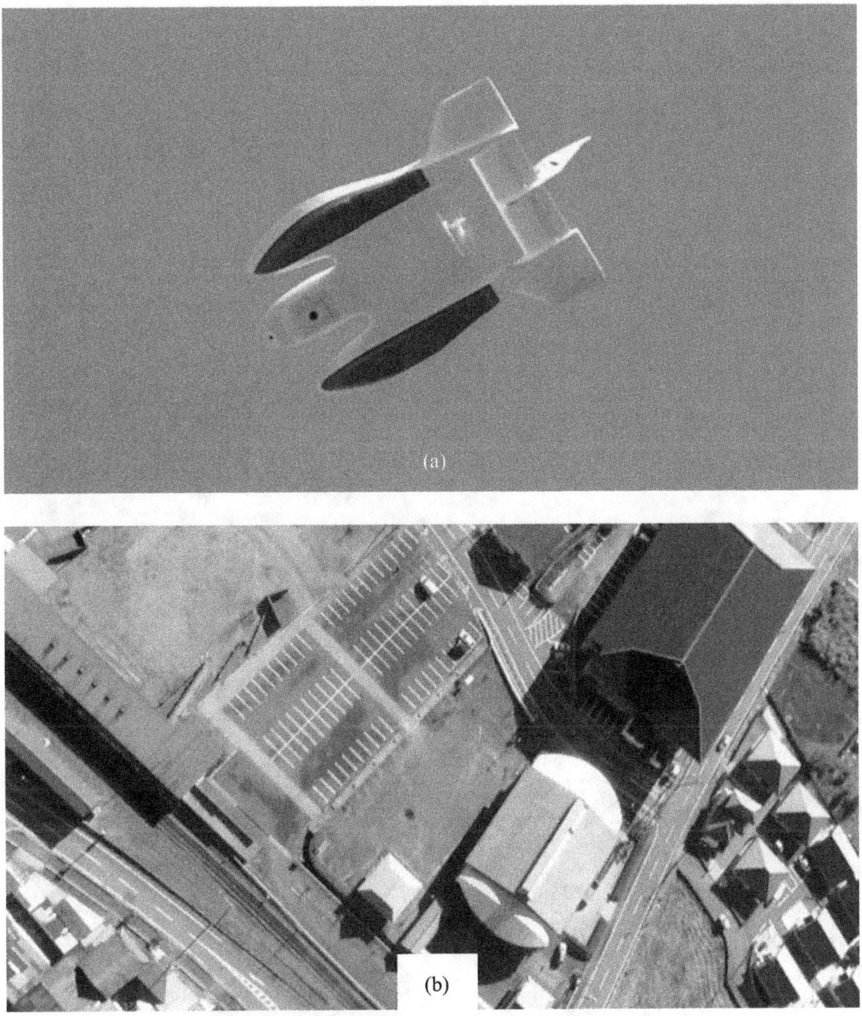

Figure 7.14 Verification experiments: (a) compact flying boat robot in flight, (b) picture from the compact flying boat robot

Table 7.2 Result of measurement

	Packet arrival rate (%)	**Radio field strength (dBm)**
Before take off	100	−67
Low altitude	86	−90
High altitude 1	88	−92
High altitude 2	80	−93
High altitude (farness)	70	−94
After landing	99	−80

Figure 7.15 Balloon-type flying robot

Figure 7.16 Pictures from the balloon-type flying robot: (a) private home and farming area, (b) river

7.1.7 Balloon-type flying robot

The balloon-type (Figure 7.15) is our new type of unmanned flying observation robot. Figure 7.16 shows the results of the verification experiments. The author and his colleagues are presently verifying the practical effectiveness of the balloon-type flying robot for disaster observation.

Figure 7.17 Example of a map of evacuation routes

7.1.8 Hazard map system

A hazard map for disaster countermeasures can be made on the basis of data obtained by an aerial vehicle system. Figure 7.17 shows an example of a map of evacuation routes. The hazard map is delivered directly to a smart phone or a tablet PC. Therefore residents can keep track of disaster information quickly and accurately. Figure 7.18 shows the unmanned aerial vehicle system for disaster countermeasures with the hazard map.

7.1.9 Conclusion

The wireless control helicopter was produced, and its payload was 3 kg; the height of the flight experiment was approximately 30 m, and it was confirmed that the observation robot can fly in a wind speed of 10 m/s. The flight time was approximately 20 min. The author and his colleagues have succeeded in the development of a system which can transfer real-time video images via the Internet while recording HD video. Video image quality of the real-time transmission is 640 × 480 pixels, 30 fps, and approximately 0.5-s delay. Figure 7.19 shows the real-time monitoring display of the system. The system can measure the levels of radiation and the amount of CO_2 in the atmosphere.

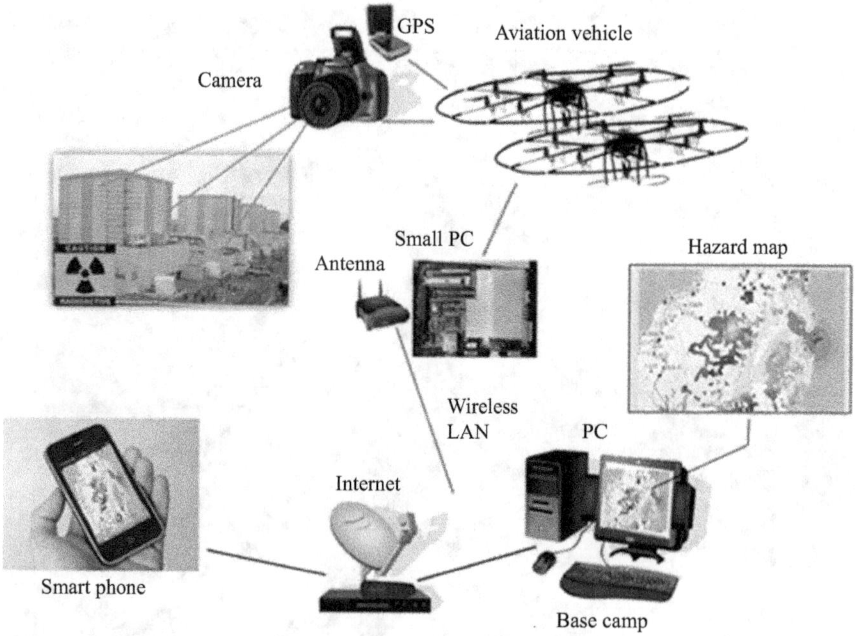

Figure 7.18 Unmanned aerial vehicle system for disaster countermeasures with the hazard map

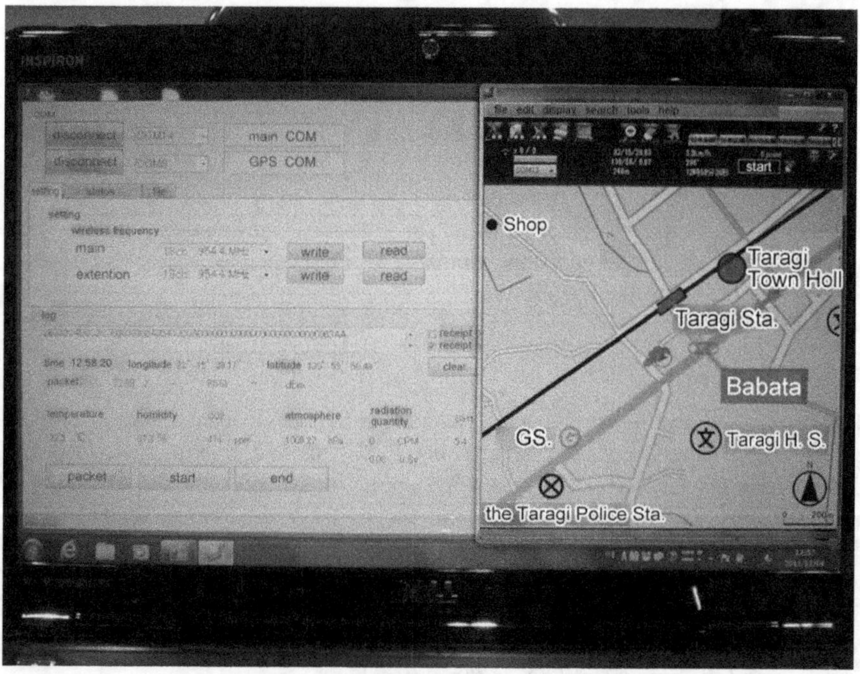

Figure 7.19 Real-time monitoring display

Furthermore, in the demonstration experiments of the disc-type flying robot, the compact flying boat robot, and the balloon-type flying robot, the author and his colleagues had positive outcomes. In the future, the author and his colleagues will advance the development of these robots to solve the issues which were identified by the demonstration experiments.

A hazard map for disaster countermeasures can be made on the basis of data obtained by the aerial vehicle system. The hazard map is delivered directly to a smart phone or tablet PC. Therefore residents can keep track of disaster information quickly and accurately. The author and his colleagues will apply the developed system to an earthquake or typhoon disaster area in the near future.

This work was supported by SCOPE, Ministry of Internal Affairs and Communications of Japan.

7.2 Space robotics

The author and his colleagues have developed a shark ray robotic fish based on biomimetic approaches. This section describes the newly developed robotic fish technology and its application to mechatronics in space. It is found that robotic fish technology creates not only new underwater robotics, but also the next generation of space mechatronics for geological survey of the moon and planets and dust cleaning in spacecraft.

7.2.1 Introduction

Living organisms' mechanisms have evolved in order to adapt to their natural environments. The evolved mechanisms have excellent manoeuvring capacities, and new robotics and machinery have often been created by implementation of biomimetic approaches. The author and his colleagues have developed robotic fish based on biomimetic approaches for the next generation mechatronics. The author and his colleagues developed the shark ray robotic fish shown in Figures 7.20 and 7.21. This robotic fish has a wing and tail structure for strong propulsion force and higher manoeuvring characteristics for turning round. First, the mechanism of the shark ray robotic fish is explained; second, its application to geological survey mechatronics is described, and finally its application as an autonomous cleaning robot for the space station is described in this section.

7.2.2 Robotic fish technology

The shark ray robotic fish has a unique structure. The shape of its head and wing are like those of a ray, and the shape of its tail is like that of a shark. It can act as stabilizer, propulsion, and rudder, as shown in Figure 6.76. The shark ray robotic fish was designed with an elastic oscillating fin system, which is applied to its tail and main wings. The robotic fish can produce a motion-induced vortex at the edge of its fin. The internal structure of the shark ray robotic fish is shown in Figure 7.22. The tail fin motors provide propulsion force and the wing motors keep motion balance. The block diagram of shark ray robotic fish control system is shown in

Figure 7.20 Shark ray robotic fish

Figure 7.21 Shark ray robotic fish swimming in water

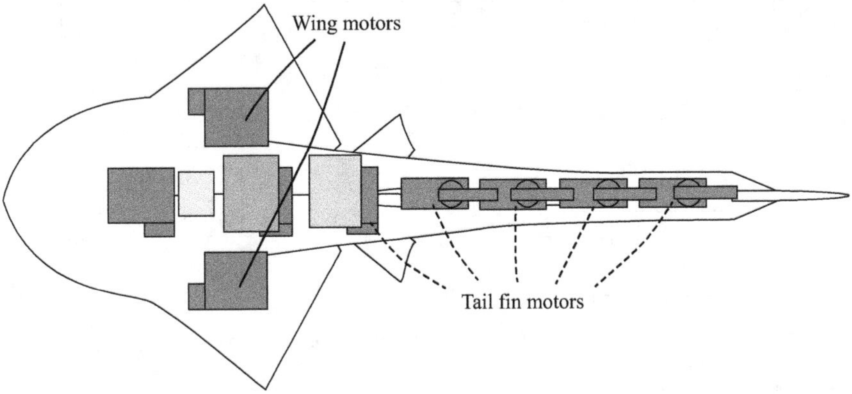

Figure 7.22 Internal structure of the shark ray robotic fish

Figure 7.23. The control is conducted by an Arduino microcontroller board, and fins are driven by a servo motor combined with the controller. The author and his colleagues experimented on thrust force for thrust performance evaluation of the robotic fish by constructing a new measuring system as shown in Figure 7.24. The system can measure the propulsion force without fixing the head of the robotic fish. The external view of the experiment system is shown in Figure 7.25. The robotic fish provides 60-N propulsion forces at 1 Hz of fin vibration frequency. If the frequency increases, a greater propulsion force can be provided. It has been shown by experiments that the structure of the robotic fish is effective for producing propulsion force.

7.2.3 Application to geological survey mechatronics

Geological surveys of the moon and planets have been proposed for developing new material resources. Figure 7.26 shows a diagram of the concept of a shark ray robotic fish surveying the geological seabed. The mechanism of the robotic fish can produce the following characteristics of a wave.

7.2.3.1 Characteristics of the wave

The vibration wave (seismic wave) consists of a primary wave and secondary wave. The primary wave is a longitudinal wave in the elastic wave that vibrates parallel to the travelling direction. The secondary wave is a transverse wave in the elastic wave that vibrates perpendicular to the direction of travel. The transmission velocity of the primary wave and secondary wave is different from that caused by the waves. The primary wave is faster than the secondary wave. Secondary waves cause small preliminary tremors. The primary wave has the characteristics of a compressional wave. A change of the volume of the medium is transmitted. Furthermore, the wave passes through a medium, such as solid, liquid, or gas. On the other hand, secondary waves cannot pass through a medium that is formless (liquid or gas), because the

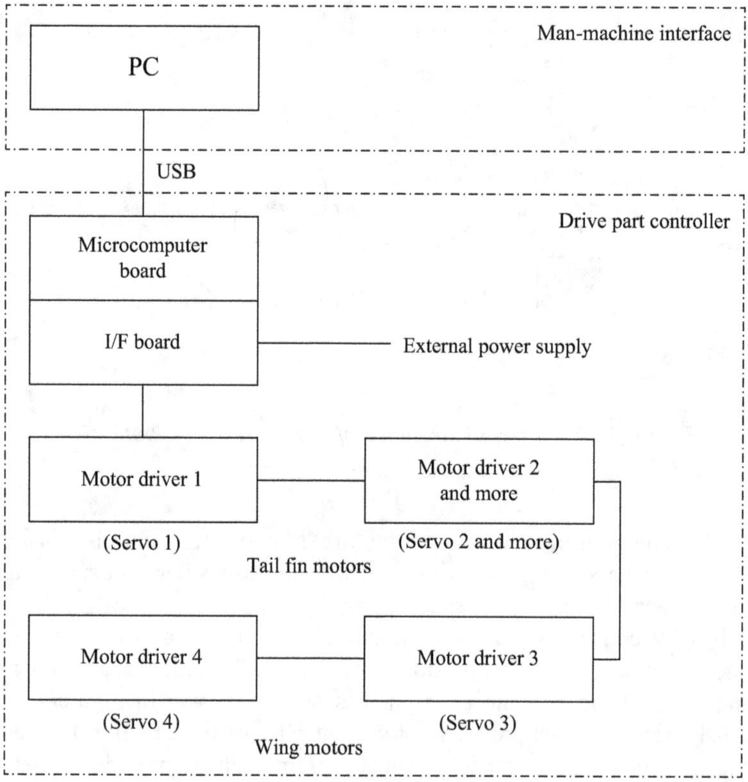

Figure 7.23 Block diagram of the shark ray robotic fish control system

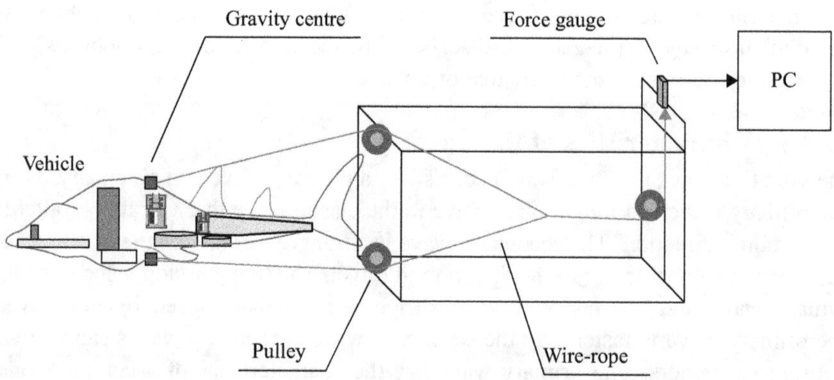

Figure 7.24 Experimental system for propulsion measurement

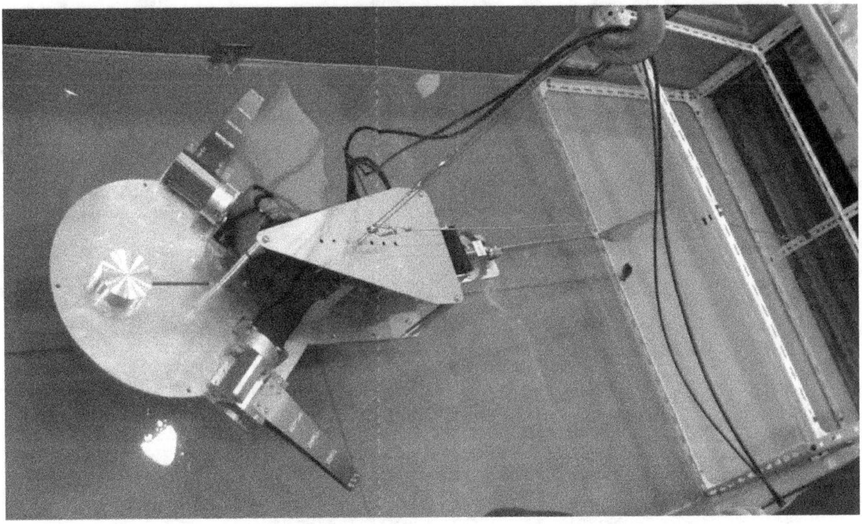

Figure 7.25 External view of the experimental system

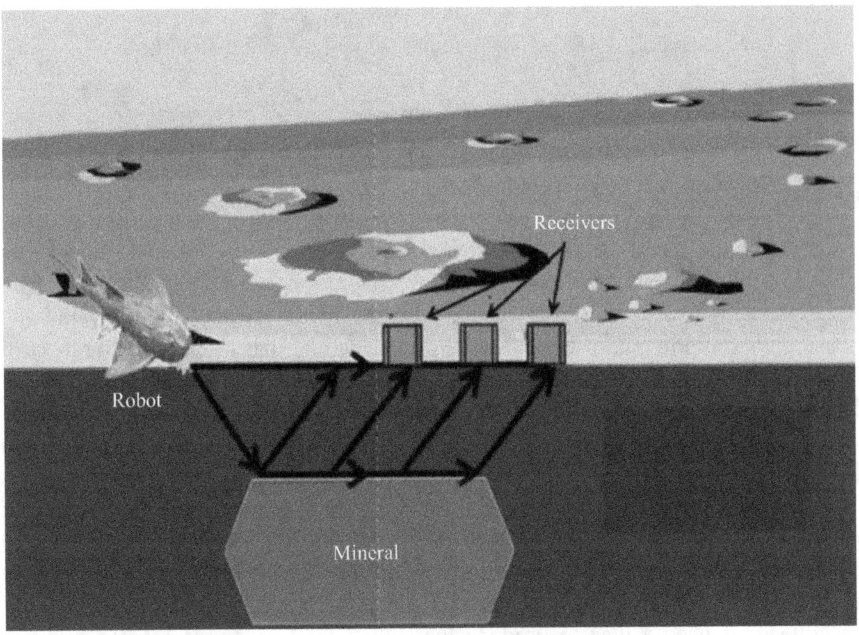

Figure 7.26 Concept of lunar geological survey

waves are transmitted by a change in the shape of the medium. Secondary waves are transmitted only through a solid medium. Currently, geological surveys are investigated using the characteristics of these waves. According to their characteristics, a geological survey method records of the waveform with a receiver.

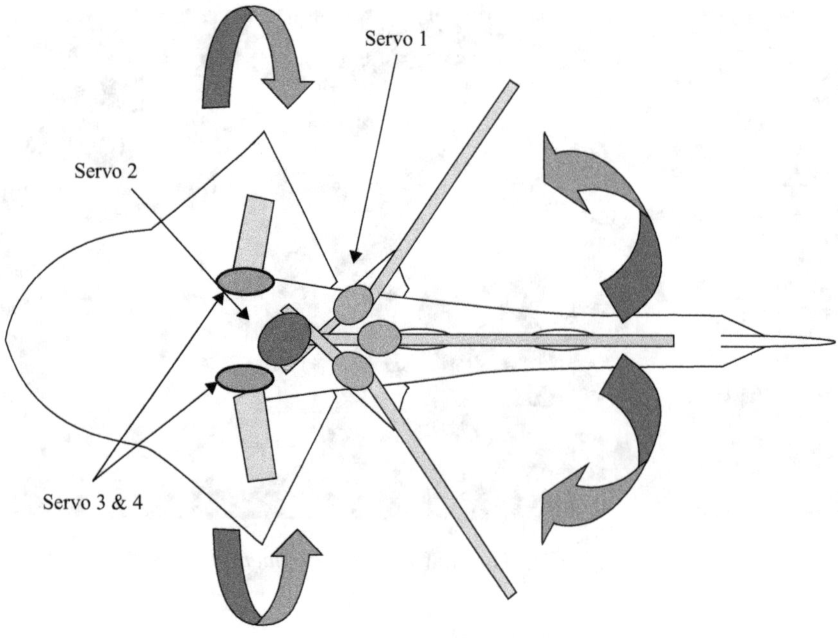

Figure 7.27 Method of primary wave and secondary wave

7.2.3.2 Geological survey method by mechatronic systems

Figure 7.26 shows the concept of a shark ray fish surveying the geological seabed. The geological survey method takes the waveform to set up a receiver to cause vibration in the shark ray robotic fish. In a geological survey, the researchers receive a plurality of locations that cause vibration waves in real time. The researchers investigate minerals that are buried in the ground by analysing the measured waveform data. Figure 7.27 shows a method that causes the vibration of the shark ray robotic fish. That causes a secondary wave that the shark ray robotic fish is adsorbed to the seafloor and moving its tail fin. In addition, it causes a primary wave, which hits the ground via pectoral fins. In this case, it is assumed that the frequency of the tail fin and pectoral fins are moving at the same time.

The author and his colleagues simulated a system to verify data measurement produced by the robotic fish. The receiver shown in Figure 7.28 is used for the experiment and can measure the vibration in three directions. Figure 7.29 shows the landscape for aligning the receiver. For the experiment, the author and his colleagues used a vice to produce the same vibration as the robotic fish and measured 10 times while shifting the interval of distance measurement.

7.2.3.3 Validation of the data

Figure 7.30 shows the measurement results. In the measurement results of Figure 7.29, the horizontal axis represents time [s], and the vertical axis represents

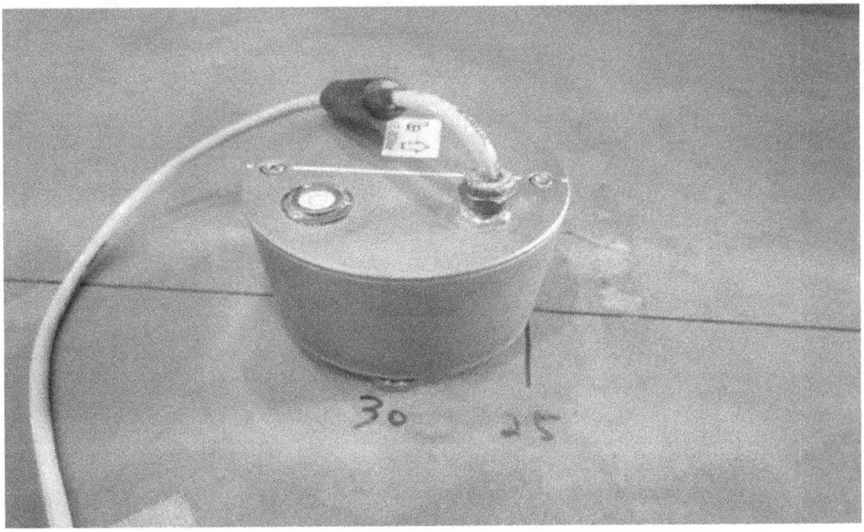

Figure 7.28 Vibration measuring instrument

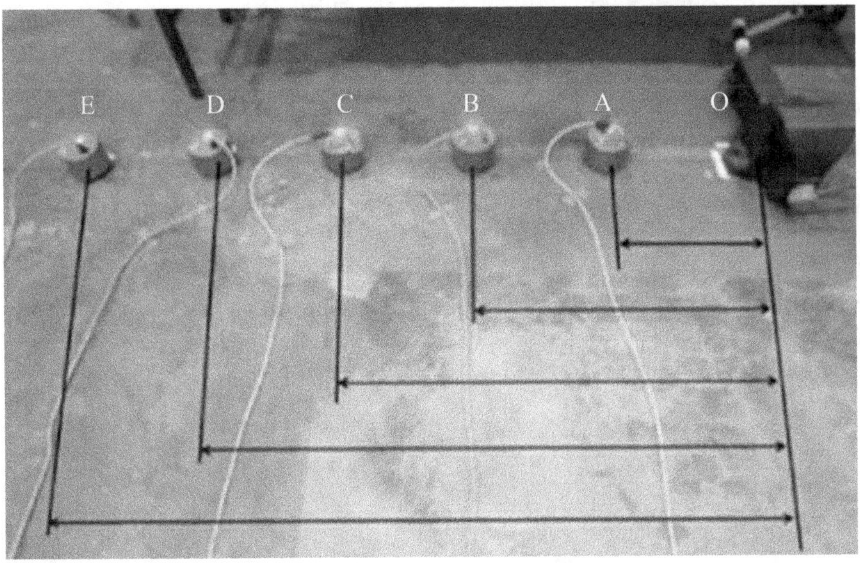

Figure 7.29 Landscape for aligning the receiver

distance [m]. The author and his colleagues raise the vibration at the position of the tO, and measure in the order of tA, tB, tC, tD, and tE. For example, the author and his colleagues consider the method of analysis of the experiment. In this experiment, the author and his colleagues have placed the five receivers, and measured the primary

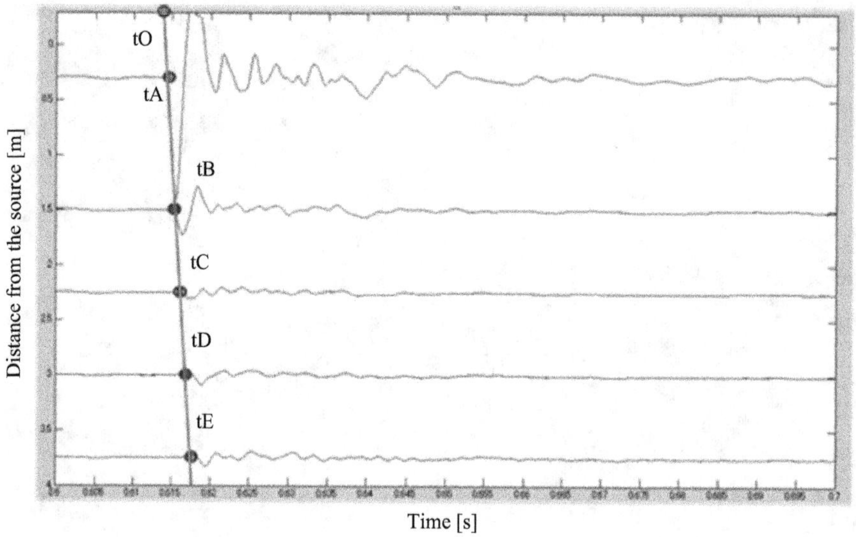

Figure 7.30 Wave vibration measurements

wave. tA, tB, tC, tD, and tE are the points where vibration begins to occur at 30 cm, 150 cm, 225 cm, 300 cm, and 375 cm, respectively, from the epicentre.

The author and his colleagues read the time at each point because the author and his colleagues do not know the time, tO, when it began to vibrate. The author and his colleagues derived the time tO using the least-square method. The author and his colleagues assumed the y-axis of time and the x-axis of distance; Figure 7.29 shows the linear alignment described by (7.1).

$$y = ax + b \tag{7.1}$$

In addition, the author and his colleagues read each point from the result that is obtained for a and b by the method of least squares. Through (7.2) and (7.3), the author and his colleagues express how to derive a and b of (7.1).

$$a = \frac{n\sum_{i=1}^{n} x_i y_i - \sum_{i=1}^{n} x_i \sum_{i=1}^{n} y_i}{n\sum_{i=1}^{n} x_i^2 - (\sum_{i-1}^{n} x_i)^2} \tag{7.2}$$

$$b = \frac{\sum_{i=1}^{n} x_i^2 \sum_{i=1}^{n} y_i - \sum_{i=1}^{n} x_i y_i \sum_{i=1}^{n} x_i}{n\sum_{i=1}^{n} x_i^2 - (\sum_{i-1}^{n} x_i)^2} \tag{7.3}$$

According to (7.2) and (7.3), the author and his colleagues can determine tO for each measurement by substituting the obtained values. Next, the author and his colleagues derive the dispersion of measurement. The author and his colleagues consider that there is dispersion due to the 10 sets of measurement data. The author and his colleagues examine the standard deviation σ to derive the dispersion. When set to time tO at the beginning of the vibration, time differences of each distance are derived (ΔtA, ΔtB, ΔtC, ΔtD, and ΔtE). For example, the author and his colleagues derive the standard deviation σA of ΔtA_i. If the author and his colleagues have n measurements, they can determine from the data ΔtA_1, ΔtA_2, $\Delta tA_3 \ldots \Delta tA_n$. Equation 7.4 shows how to obtain the average value mA.

$$\mathrm{mA} = \frac{\Delta \mathrm{tA}_1 + \Delta \mathrm{tA}_2 + \Delta \mathrm{tA}_3 + \cdots \Delta \mathrm{tA}_n}{\mathrm{n}} \tag{7.4}$$

Equation 7.5 shows how to determine the standard deviation σA from (7.4).

$$\sigma \mathrm{A} = \sqrt{\frac{(\Delta \mathrm{tA}_1 - \mathrm{mA})^2 + (\Delta \mathrm{tA}_2 - \mathrm{mA})^2 + \cdots (\Delta \mathrm{tA}_n - \mathrm{mA})^2}{\mathrm{n} - 1}} \tag{7.5}$$

The author and his colleagues investigated the change in waveform by comparing the values obtained. In addition, if a mineral is buried underground, the waveform considered to be changed is seen in the effect of the reflected wave. The author and his colleagues derive the velocity of the primary wave and secondary wave from these calculation methods. In addition, Poisson's ratio ν is derived for the stratum structure to examine something. This is an indicator of the strain of the formation. Further, it is possible to find something in their formation. Equation (7.6) shows how to calculate Poisson's ratio.

$$\nu = \frac{(V_p/V_s)^2 - 2}{2\{(V_p/V_s)^2 - 1\}} \tag{7.6}$$

In this experiment, a value $\nu = 0.2$ of Poisson's ratio from the calculation results were obtained. Referring to the mechanical properties, the value of the Poisson's ratio is concrete.

7.2.4 Application to a space station cleaning robot

The Environmental Control and Life Support System (ECLSS) controls temperature and humidity, oxygen and carbon dioxide concentration, air pressure, fire extinguishing, and other attributes in the International Space Station (ISS) cabin. Dust is a problem in the closed cabin environment. Dust in the cabin is adsorbed in the air conditioning filter; however, it is not perfect. Astronauts use a vacuum cleaner to clean dust that cannot be absorbed by air conditioning. Reducing the load of astronaut cleaning is a challenge. A method for reducing their cleaning work is required for increasing the efficiency of the astronauts' mission. In this study, the author and his colleagues chose fish shapes for the cleaning robot. A creature suitable for movement under water has the characteristics of a streamlined body and fin for propulsion and attitude control. The fish shape is excellent for transferring efficiency and manoeuvrability in three-dimensional space. Elastic oscillation the fin was realized to have the high controllability and soft behaviour of creatures. A robotic fish is safe, because the robot's exterior is not equipped with rotary parts, such as propellers. This section describes the development of autonomous robot to clean dust in the ISS.

7.2.4.1 Overview of robotics fish

In this study, the author and his colleagues choose fish shapes for the cleaning robot shown in Figure 7.31, using a fin to move freely.

An air bladder is not required for movement in zero gravity. There are many delicate precision instruments in the ISS cabin. Therefore, detecting and avoiding collisions requires an autonomous control and manoeuvrability system. In addition, the robotic fish must not harm the astronauts. Even if a collision is detected, robot must have a safe shape. For this reason, the robot is not equipped with external rotary parts. The robotic fish takes evasive action by detecting the astronauts, walls, and obstacles. The cleaning robot having autonomous flight control and navigation is included in the Japanese Experiment Module named *kibo*. It has a function to collect dust which cannot be recovered by the ISS air conditioning system. The robotic fish starts using autonomous navigation within a few seconds of turning on the power. Then, autonomous navigation avoids astronauts and walls. Recovering dust from the robotic fish is conducted by a vacuum cleaner used by the astronauts. It was designed like a round whale shark, which orders to calm down for astronauts. Small fish and plankton are swallowed by a whale shark's wide open mouth. Likewise, dust is sucked in by the robot fish's wide open mouth. The body size is limited by regulations to less than 200 mm for transport by rocket. Therefore, it requires a

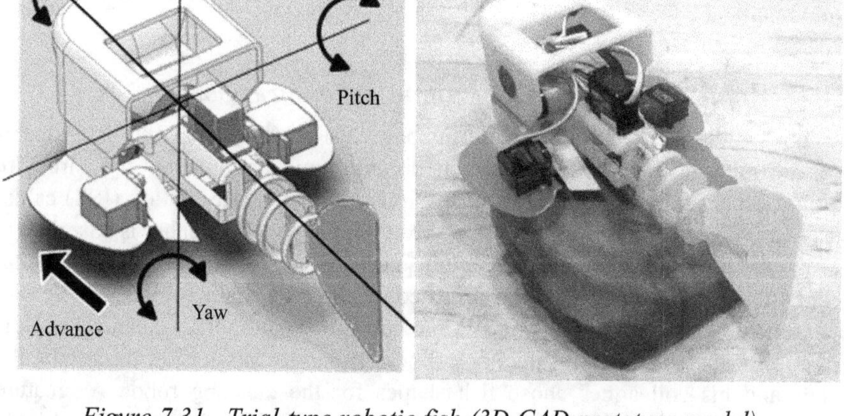

Figure 7.31 Trial-type robotic fish (3D CAD prototype model)

very compact movement and control mechanism. The size of the robotic fish is 199 mm × 135 mm × 71 mm, with a total weight of approximately 0.2 kg.

The outer shell of the robotic fish was built by a 3D printer, and the autonomous navigation algorithm was developed and installed in the controller.

7.2.4.2 Thrust and dust cleaning mechanism

Figure 7.32 represents the front view of the robotic fish. Main thrust is generated in two propellers to turn counter-clockwise. The rotations of the propellers countervail each other's rotation moment. Air is aspirated by the large opening in front of the robotic fish, and breathed out to the steering mechanism in the rear. A dust filter is equipped in front of the propellers. While cleaning and performing inertial navigation, a small negative pressure generated in the large opening in the front of the robotic fish. Therefore, the propeller thrust also acts to collect dust. The dust filter is applied to absorb dust coating.

7.2.4.3 Steering mechanism

Attitude control uses the air breathed out from the rear of the robotic fish, with bilateral pectoral fins and a caudal fin. Attitude control for the roll, pitch, and yaw axis is possible by the steering mechanism. It is an efficient method; expelled air is directly exposed to the steering mechanism. The elastic oscillation fin allows for a large curvature, which enables a quick yaw axis change. The elastic oscillation fin generates a small thrust from fin vibration. Fin action advances by shaking the body like a fish, and is effective in searching for an obstacle. To stop, the

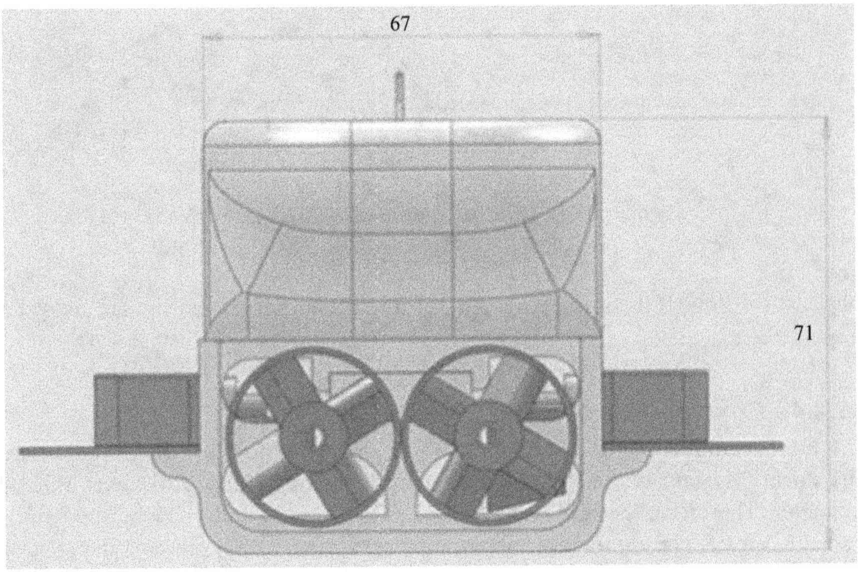

Figure 7.32 Front view of the robotic fish

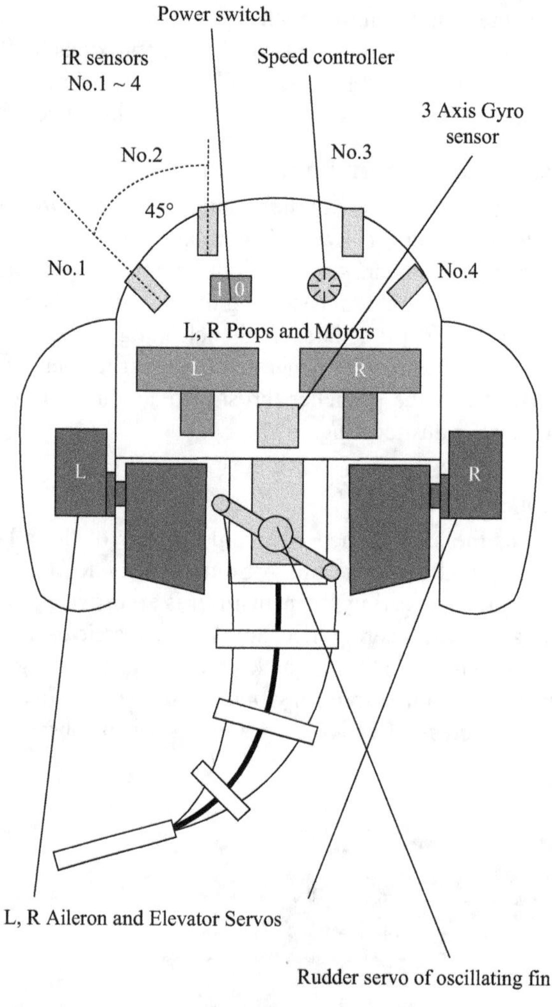

Figure 7.33 Control system of the robotic fish

robotic fish turns 180° on the yaw axis by the elastic vibration fin, and uses the propeller thrust.

7.2.4.4 Control system

The ISS is equipped with LR6 batteries that can supply power to the robotic fish. The control system of the robotic fish needs a compact and low-power microcomputer. Therefore, the Arduino Pro Mini 328 (size: 33 mm × 18 mm × 3 mm; Vcc: 3.3 V) was chosen for the microcomputer. Figure 7.33 represents the control system of the robotic fish. Starting and stopping the robotic fish is performed by a power switch. The flight speed of the robotic fish is controlled by a volume controller. The flight speed can be set to 100 mm/s or less. Obstacles are detected by four infrared sensors, which enable the robot to perform evasive action. The sensor arrangement detects the direction and distance of an obstacle with respect to the travelling direction. The obstacle direction is detected by an infrared sensor, and the movement is adjusted according to the autonomous navigation algorithm. When the robotic fish detects a collision with astronauts and obstacles, it goes into a spin state. Therefore, a three-axis gyro sensor is used to detect the angular acceleration. The robotic fish recovers from the spin state by its fins, and returns to autonomous navigation. The spin recovery operation is the highest priority in autonomous navigation.

This section has described a trial product of an autonomous robotic fish to clean dust in the ISS. A 3D printer can quickly manufacture parts that are designed in 3D CAD. The elastic oscillation fin was realized to mimic the high controllability and soft behaviour of creatures. It was predicted that the simple algorithm of autonomous navigation can cruise the cabin almost equally to astronauts. In the future, the robotic fish is required to undergo an electro-magnetic compatibility test, vibration test, aging test, and off gas test.

7.2.5 Conclusion

The author and his colleagues have developed the robotic fish. The robotic fish technology can create new underwater robotics which can perform life-like swimming. It is found that the technologies of the robotic fish have much potential for creating new space mechatronics through these applications. The author and his colleagues still continue to conduct further experiments to produce these space products. The author would like to express his sincere gratitude to NASA, JAXA, and Dr Masahiko Mizui and colleagues for supporting the research.

Chapter 8
Advances in medical robotics and mechatronics

8.1 Surgical robotics and mechatronics by biomechanical technology

8.1.1 Introduction

In a surgical operation, the research and development of a surgical robot that is both safe and effective is needed in consideration of decreasing the physical and mental strain of the surgeon and the patient. By observing the motion of fish, it has been found that fish swim efficiently using the flexibility of pliable fins, giving the actuator the name of 'flexible oscillating fin propulsion system'. The author proposes the application of the elastic vibration wing propulsion system to the realm of healing. A flexible forceps robot was developed for surgical operations with a vibrating flexible fin propulsion system application.

In particular, for surgical operations, the technology can decrease physical and mental strain of the surgeon and the patient. Thus it is necessary to research and develop a surgical robot that is both safe and effective.

It is well known that marine creatures such as fish swim using little power even at high speeds (dolphin: 60 km/h; swordfish: 80 km/h) and sweetfish and others are superior in their resting state maintenance characteristics. Through the observation of the motion of fish, it has been found that fish swim efficiently using the flexibility of their pliable fins.

In this study, the author and his colleagues developed a robotic forceps for surgical operation by applying a flexible oscillating wing propulsion system.

8.1.2 Drive principle of flexible oscillating fin propulsion system

The development of the robotic fish started with the development of an actuator that could realize fishlike fin movement. The author named the actuator a 'flexible oscillating fin propulsion system'.

Figure 8.1 shows the drive principle of the early flexible oscillating fin propulsion system. The flexure like a fish fin was produced by variously changing elastic modules of the flexible oscillating fin. The optimal shape of the fin was also

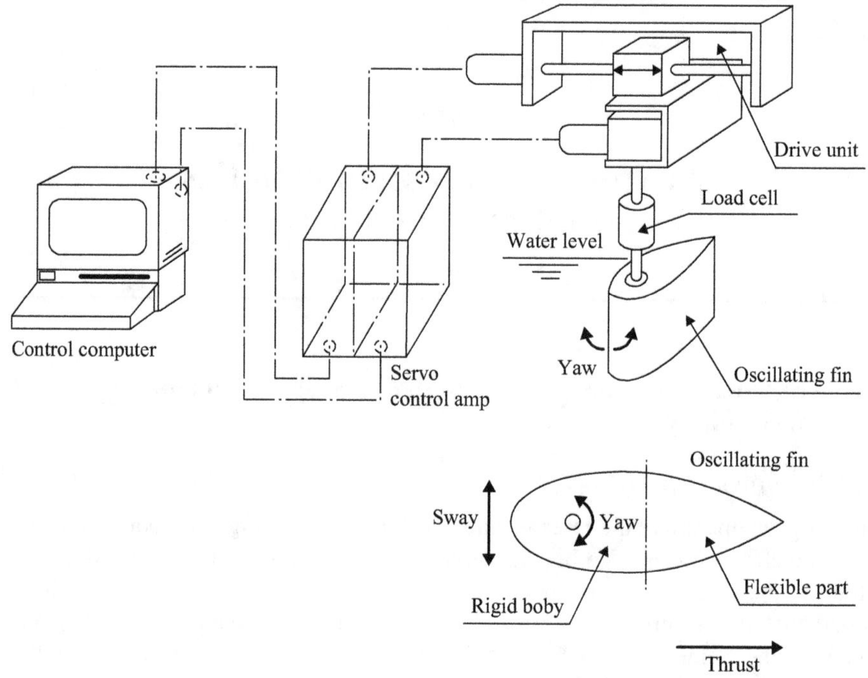

Figure 8.1 Drive principle of flexible oscillating fin propulsion system

chosen through numerical simulation and tank testing. Then, the manoeuvrability was tested with a small craft, on which the oscillating fin propulsion system was installed.

8.1.3 Application of robotic fish technology

On the basis of the basic technical research of the flexible oscillating fin propulsion system, an artificial fish has been developed as an applied product of this system. The first developed robotic fish, a sea bream, has gained a reputation in the world for swimming just like the real thing without power and communication cables. The robotic fish has a propulsion mechanism with a flexible/rigid oscillating fin, and can swim in water quietly, as shown in Figure 8.2. The flow of the surrounding fluid when the fish swims was made visible with the robot. The Kalman vortex can be observed backward, as shown in Figure 6.68.

The development model of the robotic fish is shown in Figures 6.65 and 6.66. It is composed of a flexible/rigid oscillating fin, actuator, battery, charge unit, hydraulic pressure sensor, sealing water pump, waste water pump, ballast tank, control unit, and wireless telecommunications unit.

The author and his colleagues have been developing many prototypes of AUVs, and studied the flapping wing type which could be used to collect oceanographic data and water samples. The manta-type robot moves a fin in the manner of a bird flapping its wings, and the fish robot swims flexibly to 'fly' in the water. The elastic vibration wing propulsion system demonstrates a large impellent and lift, and achieves a high performance. Its outline resembles the *manta birostris* (manta ray) as shown in Figures 8.3–8.5.

Figure 8.2 *Robotic fish swimming in water*

Figure 8.3 *Flapping flat fish*

Figure 8.4 Somersault flat fish

Figure 8.5 Turning flat fish

Figure 8.6 Structure of flapping wing

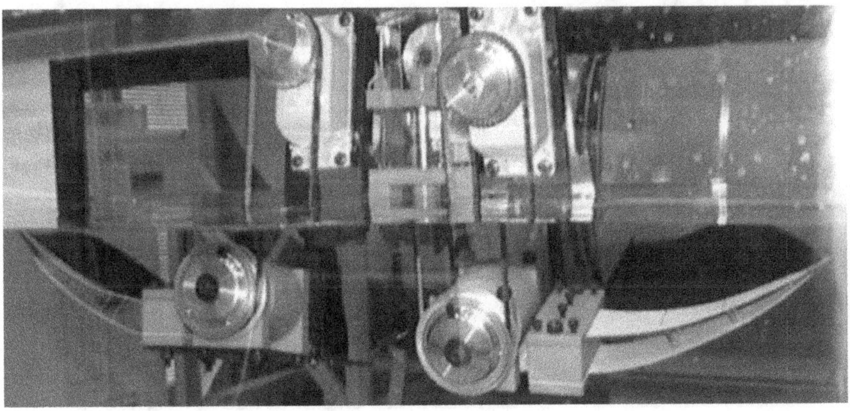

Figure 8.7 Flapping wing with multiple joints system

The multiple-joint propulsion system has a twisting mechanism to realize higher manoeuvrability of the underwater vehicle as shown in Figures 8.6 and 8.7.

8.1.4 Development of a flexible forceps robot for surgical operation

For the development of the forceps robot by the finite element analysis method, the forceps robot is modelled and analysed. The author and his colleagues developed the forceps robot so that when operating on the intestines, the forceps robot is used to hold internal organs.

The prototype gripper of the forceps robot was modelled as shown in Figure 8.8. The material of the forceps is SUS304 and the main characteristics of the small intestine are shown in Table 8.1.

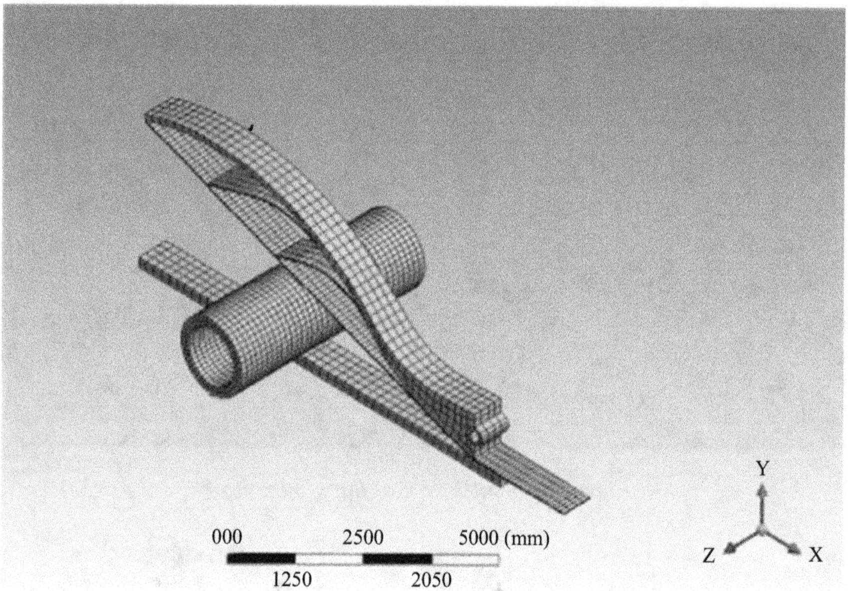

Figure 8.8 Prototype model

Table 8.1 Mechanical properties

Properties / Materials	Young's modulus E (MPa)	Poisson's ratio σ
Sus304	1.95×10^5	0.25
Small intestine	3.16	0.475

8.1.5 Production of forceps

The new forceps has the function that lends itself to the elastic vibration wing system and the regulating system, and can correspond to patient's blood pressure variation under the operation. The forceps robot system with this function was made for trial purposes in the present study. Figures 8.9 and 8.10 show the entire system and the gripper.

When the blood pressure rises, the holding pressure must be increased to prevent the content leaking; when the blood pressure decreases, the holding pressure must decrease the load of the organization and internal organs. This regulating system is made possible in this research.

8.1.6 Comparison of conventional forceps and new forceps

As shown in Figure 8.11, for the new forceps and previous forceps, the author and his colleagues measured holding pressure and holding force (pull force); compared

Figure 8.9 Whole system

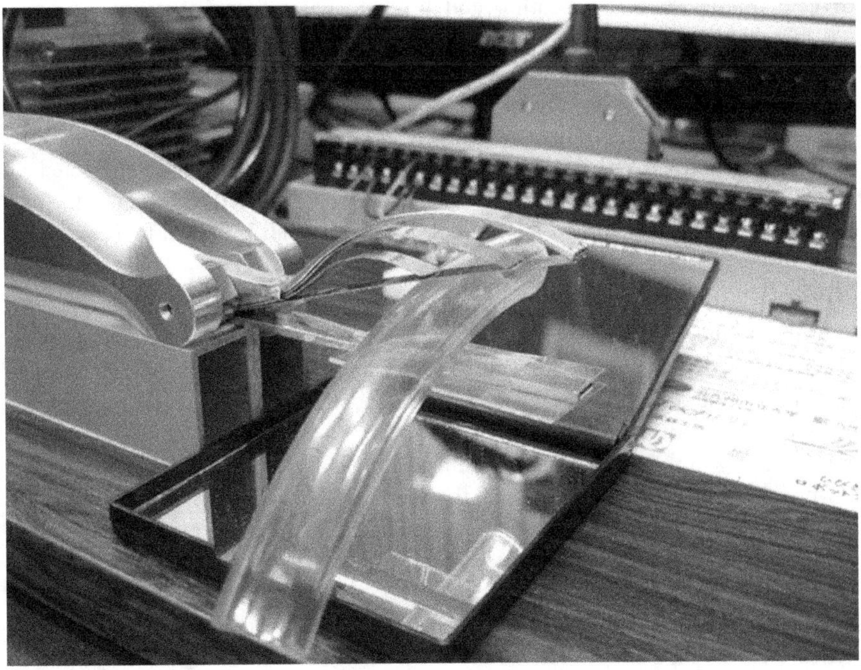

Figure 8.10 New forceps

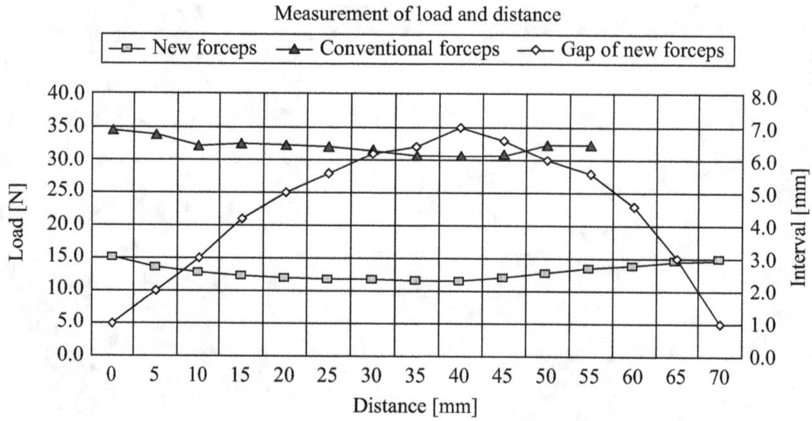

Figure 8.11 Comparison of performance

with the previous one, the author and his colleagues can fully prove that the hold pressure of the new forceps was less than 60% of the previous one (average: previous forceps, 32.2 N; new forceps, 12.9 N). There is a 7-mm grip in the centre of the new forceps and maintain a constant gripping force, even when it is used in a surgical operation.

According to Figure 8.11, we could prove that the holding force was the smallest in the middle of the holding area of the new forceps. The experiment is shown in Figure 8.12. The author and his colleagues found that a pig's small intestine that is similar to a human's small intestine performing the holding experiment showed that the middle of the new forceps can be utilized.

A medical report showed that pressure in small intestines was 0.53 × 105 Pa (40 mmHg) or more (approximately 544-mm H_2O); this experiment poured water from the position of 550 mm in height into small intestines. Pressure acted on the turning angle θ and on the contact side in the gripper when the gripper of the new forceps was shut, and water which did not leak. The result is shown in Table 8.2.

8.1.7 Conclusion

According to the finite element method analysis, with the new forceps as shown in Figure 8.13, it is possible to hold for making the power distribution average and wrapping soft internal organs. Power can be concentrated on the hand tip, and it is not necessary to powerfully hold and pull out.

According to the data of the experiment, the analysed and obtained results are shown in Figure 8.14. On the Y-axis, the stress is about 4.26 N. The error margin caused a microindentation of the actual small intestines that could not be reproduced well by CAD.

From the results of the research, the author and his colleagues can develop the new forceps that can correspond to the shape of held internal organs, and fix the internal organs with a shape that can be easily changed with soft and average power.

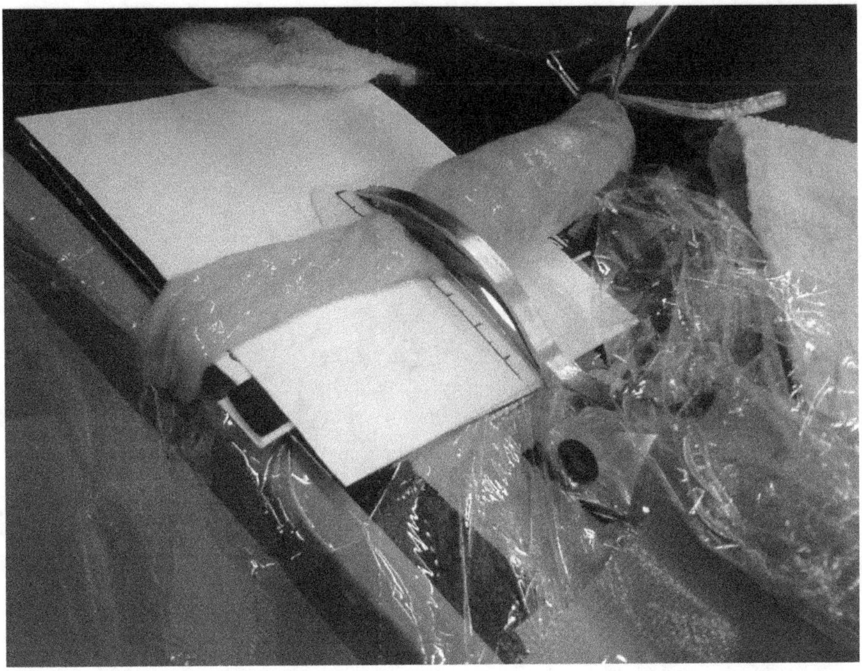

Figure 8.12 Grip force measurement

Table 8.2 Specification of the prototype model

Rotation θ (°)	Normal stress F (N)	Contact area S (mm^2)
4.3	4.46	16.1

8.1.8 Further approaches of surgical instrument development

8.1.8.1 Seamless development of surgical instruments based on biological mechanisms using CAD and a 3-D printer

Introduction

In the area of manufacturing surgical instruments, the ability to rapidly design, prototype, and test surgical instruments is critical. This section provides a simple case study of the rapid development of two biomechanism-based surgical instruments which are ergonomic and aesthetic, and were successfully designed, prototyped, and conceptually tested in a short period of time.

Recently, there has been an increasing need for surgical instruments that can hold organs delicately yet stably. Such an instrument increases the efficacy of surgical operations by decreasing the physical and mental strain on both surgeons and patients. New biomechanism-based surgical instruments, based in part on the anatomical structure of a fish, provide soft-handling forceps where pressure is

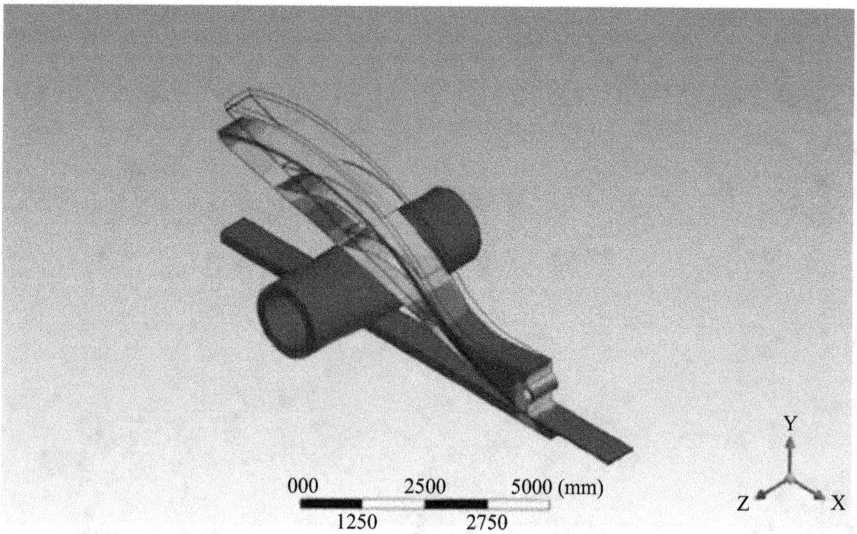

Figure 8.13 Equivalent stress distribution

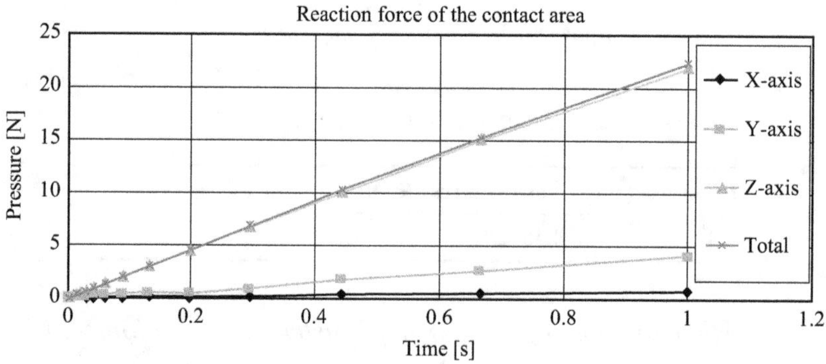

Figure 8.14 Measurement of load and distance

distributed over a larger area. The author and his colleagues have used a seamless design and prototyping process. This process has been used to prototype biologically based mechanisms using 3-D CAD and a 3-D printer. Specifically, a fish-based mechanism which produces an elastic oscillating fin and shark skin which effectively reduces hydrodynamic resistance have been found to be effective in creating superior surgical instruments. Classical surgical instruments, which exert a large amount of force on a limited number of points on organs, are stressful for both surgeons and patients. Furthermore, classical instrument ergonomics often do not consider the surgeon and are often awkward and thus stressful to use.

The purpose of this research is to develop user-friendly surgical instruments that enable more efficient surgery. This process is effectively facilitated using a more seamless design through to the prototyping process using 3-D CAD and a 3-D printer. The surgical instrument's design and concept prototyping are discussed below.

Surgical instrument design

The basic steps of the system design and prototyping are shown in Figure 8.15. First, the specifications, such as a surgical instrument's function, form, size, elasticity, and material characteristics, are designed to suit the needs of the specific instrument. Specifications of some parts may mean creating a new kind of engineering. For the kinetics of surgical instruments, motion simulation is carried out in CAD. Next, verification of the resultant measurement data of the surgical instrument is carried out. In considering the design of medical instruments, aesthetics, ergonomics, and functional efficacy must be considered. Techniques for data measurement of the design include strength modelling based on the target production materials. This process uses STL 3-D data which may be directly sent to a 3-D printer, thus making

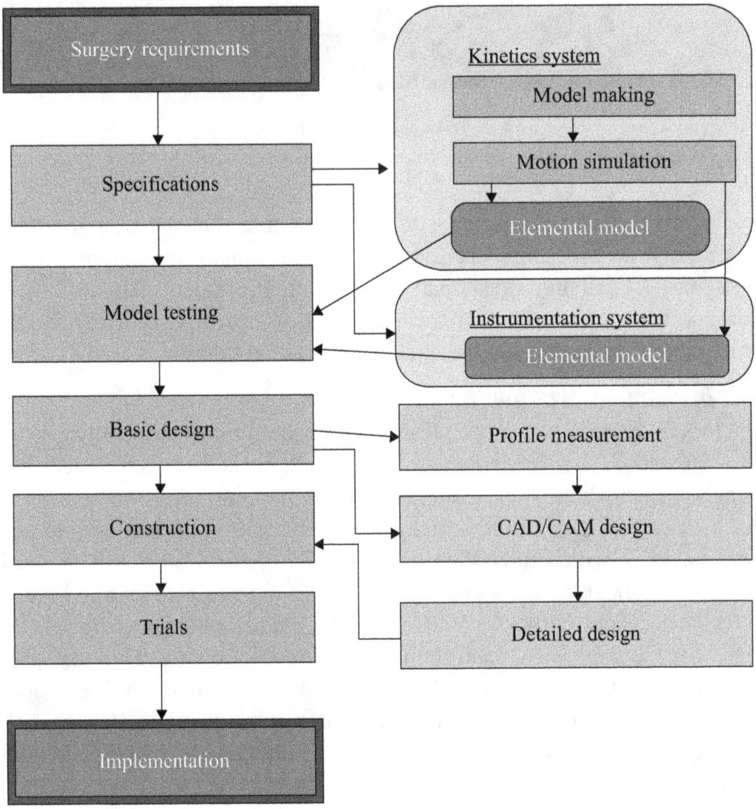

Figure 8.15 Basic steps of design and prototyping

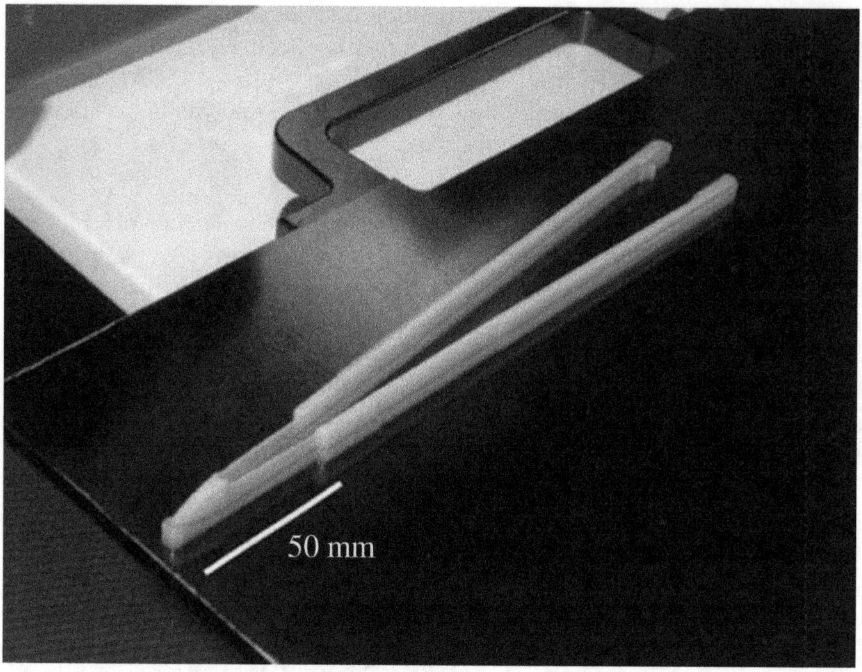

Figure 8.16 3-D printed extractor

the rapid prototyping of surgical instruments possible. On the basis of the basic plans, assembly drawings are created. Complex mechanisms can be reduced to individual parts. Each individual part making up the mechanism is prototyped. Finally a surgical instrument is created. Prototyped instruments undergo practical tests, and, if the functionality is satisfactory, the process ends.

Surgical instrument prototyping

The developed surgical instruments based on these biomechanisms are shown in Figures 8.16 and 8.17. The length of the extractors is 295 mm. Regarding the extractors, the instrument's shark skin-based soft grasping mechanism makes them easy to use, allowing a surgeon to operate smoothly despite the instrument becoming covered with fluids such as blood. The extractors' grip is stable on account of the shark skin-based surface. This set of extractors shown in Figure 8.16 was prototyped in approximately 30 min. The STL data was sent to the 3-D printer, an Objet 260 Connex with maximum resolution of 16 μm. The soft grasping forceps are shown in Figures 8.17 and 8.18. It is estimated that the time required to manufacture this surgical instrument would be approximately 40 h using CNC milling. The feedback to design is conducted rapidly. This allows for iterations as required prior to production of the final concept prototype.

Table 8.3 shows the specifications of the forceps; both target specifications and those achieved to date are shown. Figure 8.19 shows an approximated stress

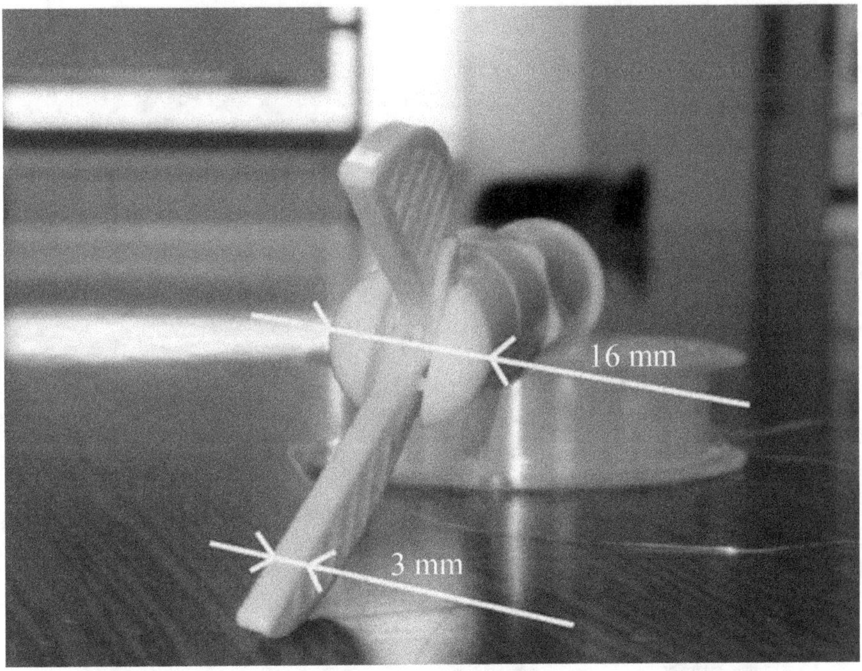

Figure 8.17 3-D printed forceps

analysis (yield) using 3-D CAD (SolidWorks simulation) based on the strength of VeroWhite PolyJet resin. It must be noted the vertical struts in the target forceps' jaws should provide a degree of freedom at both ends to provide for an equal distribution of pressure along the jaw's surface, in order to exert even pressure on the organ being handled. This mechanism shown in Figure 8.19 reflects the design of the flexible oscillating fin used for the robotic fish. The prototyped product is shown in Figure 8.17; its production time is approximately 2.5 h. The products are made with VeroWhite PolyJet resin material. It must be noted that for this particular application, the VeroWhite PolyJet resin lacked the tensile strength to provide fully functional medical instruments; however they enabled confirmation of the concept. In this regard a 3-D printer capable of printing materials of higher tensile properties will be needed, such as powder-based laser sintering (direct metal) or more traditional CNC milling technology. In the case of the forceps, the strength of the printed resin is sufficient for larger or thicker parts; however, for thinner parts or parts providing intricate mechanical functionality, parts with metallic strength will be required. For this experiment, in the case of the forceps the size of intricate mechanical parts were scaled up to confirm the mechanical feasibility of the mechanism; hence the two sets of data in Figure 8.15, 'Target' and 'Experimental'. Other fundamental limitations include the maximum printable length (diagonal) and the ability of be able to physically remove the support material (photopolymer) produced by the 3-D printer.

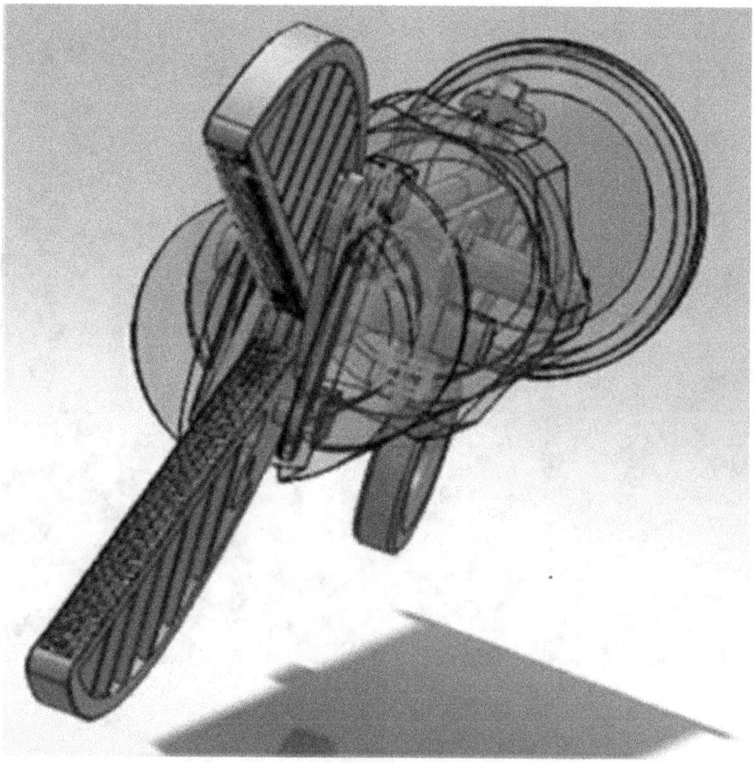

Figure 8.18 CAD design of the forceps

Table 8.3 Specification of the forceps

Parameter	Prototype model
Total length	130 mm
Hand grasp section length	45 mm
Tip length	50 mm
Tip thickness	3 mm
Overall tip thickness	16 mm
Opening angle (max)	45°

The author and his colleagues have confirmed a seamless workflow from design through to conceptual prototype. An experimental test of the product is conducted shown in Figure 8.20. In addition, the developed elastic biological forceps is summarized in Figure 8.21, and experimental results between the normal (rigid) and elastic structure are shown in Table 8.1. The elastic effectiveness for soft handling (smaller average pressure) is verified in several loads. Application of biomechanism to the forceps is summarized in Figure 8.22.

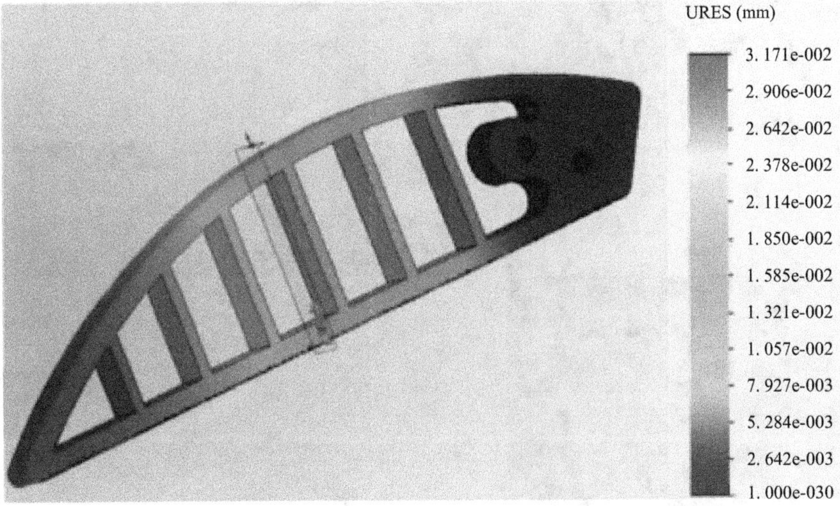

Figure 8.19 Stress analysis of the forceps jaw

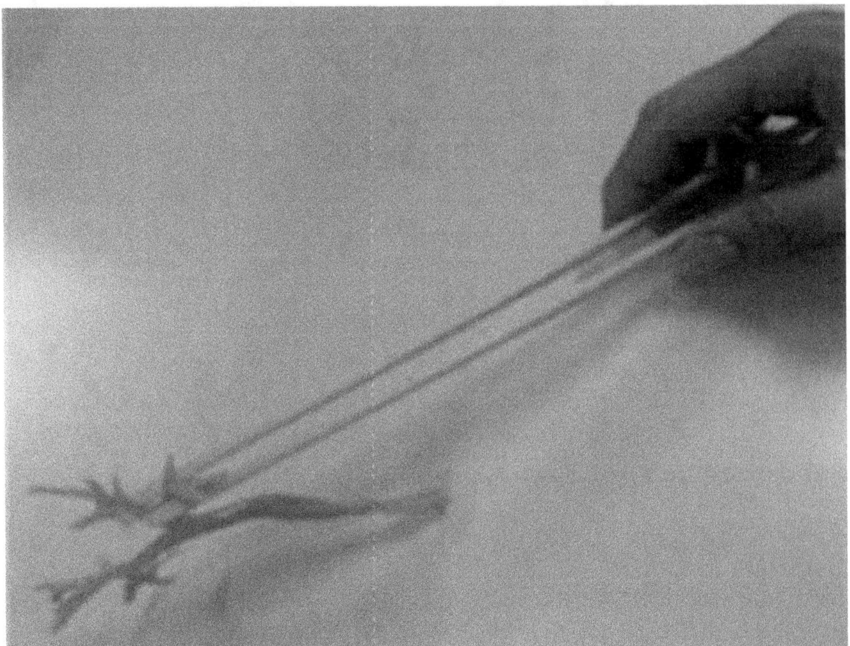

Figure 8.20 Experimental test to grasp an organ with 3-D printed forceps

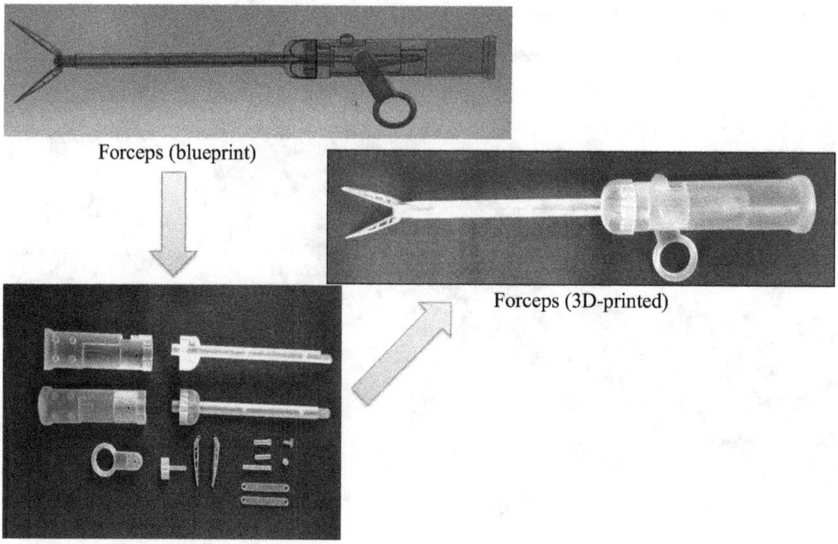

Figure 8.21 Elastic vibration wing forceps

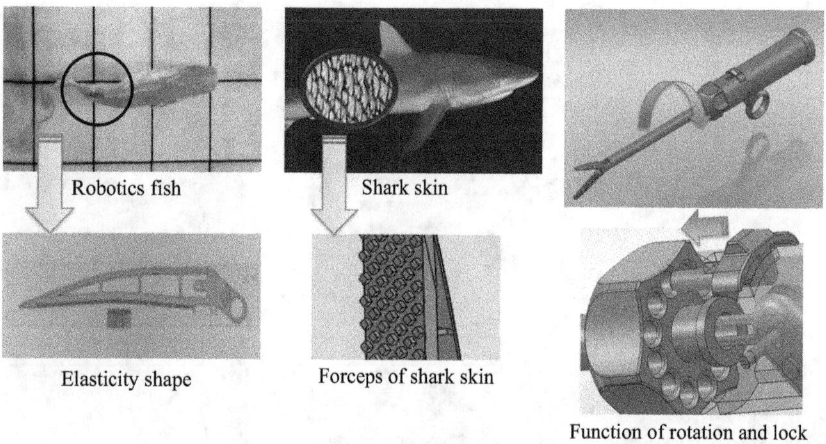

Figure 8.22 Application of biomechanism to the forceps

Conclusion

A new basic method of seamless design through prototyping has been utilized, which can significantly reduce time to production. In this study, on the basis of a biomechanism, an extractor and forceps which are ergonomic and aesthetic were successfully designed, prototyped and conceptually tested in a very short span of time.

In order to prototype fully functional instruments, because of the limited strength of the 3-D printer resin, more traditional manufacturing such as CNC

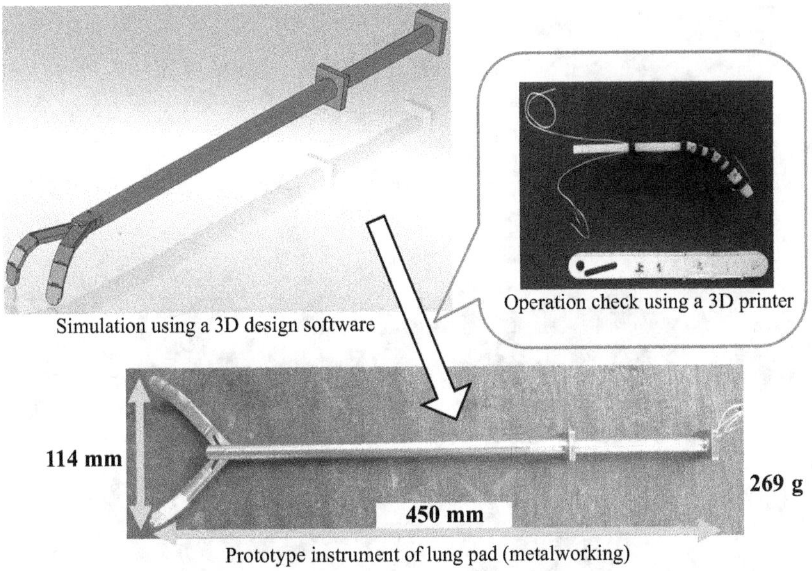

Figure 8.23 Prototype instrument of lung pad

milling or powder-based laser sintering (direct metal) would be required particularly in the case of thin-walled parts or parts providing intricate mechanical functionality.

The author and his colleagues plan to make other surgical instruments, specifically, lung forceps based on a bioflexible mechanism using this process. A prototype instrument of lung pads by a 3-D printer and finally by metalworking are shown in Figure 8.23. The 3-D printer is set up at the Hybrid Medic Training Centre in the Nagasaki University Hospital; this enables surgeons to provide immediate feedback to engineering researchers by promptly evaluating the developed experimental production instruments.

A part of this study was supported by the Medical Engineering Hybrid professional development program staff. The author and his colleagues would like to express their sincere gratitude to Nagasaki University, University of Occupational and Environmental Health, and related personnel in this research and development.

8.2 Rehabilitation robotics and mechatronics

8.2.1 Introduction

The author and his colleagues have developed a wrist rehabilitation robot for hemiplegic patients. The robot is effective for repetitive rehabilitation, and useful for patients. A mirror effect control system is developed on the basis of the experience that a patient can easily move a limb while moving a limb on the other side. The effectiveness is successfully confirmed by practical testing in a hospital.

The effective wrist rehabilitation robot shown in Figure 8.24 has been developed at practical level for hemiplegic patients, who need repetitive rehabilitation training

Figure 8.24 Effective wrist rehabilitation robot

to improve the function of upper and lower extremities and to increase the range of motion. The effectiveness of repetitive rehabilitation training is shown in medical studies. The mildly physically disabled patients can undergo self-training with such rehabilitation tools as blocks. In contrast to this, the massively physically disabled patients generally undergo forced training with training support devices. However, forced training fails to be more effective because patients cannot exercise on their own active mind. Therefore, the effect of the treatment is unable to surpass the critical limit. Although occupational therapists can give the patients undivided attention, the number of occupational therapists is limited and there is a compelling need to reduce their work load. Many research results for rehabilitative training robots have been reported; however, it is often the case that the robot system is large and the process is complicated. Therefore, it is difficult for patients to feel unrestrained to use the robot system. Thus, the research is aimed at developing a rehabilitative robot that enables the effective training of the patient. The training robot conducts several modes of training. Specially, the paretic hand can undergo the rehabilitative training by the other non-paretic hand based on a mirror effect control system.

8.2.2 System configuration

The developed rehabilitative robot consists of passive, constant torque and autonomous training modes. In addition, the author and his colleagues have developed a training robot that has a function to detect the myoelectric potential by myoelectric sensors on the patient's muscle, and analyse the patient's intention to flex the wrist as shown in Figures 8.25–8.27. The training robot can provoke training through the function, so that patients can undergo effective training as shown in Figure 8.28.

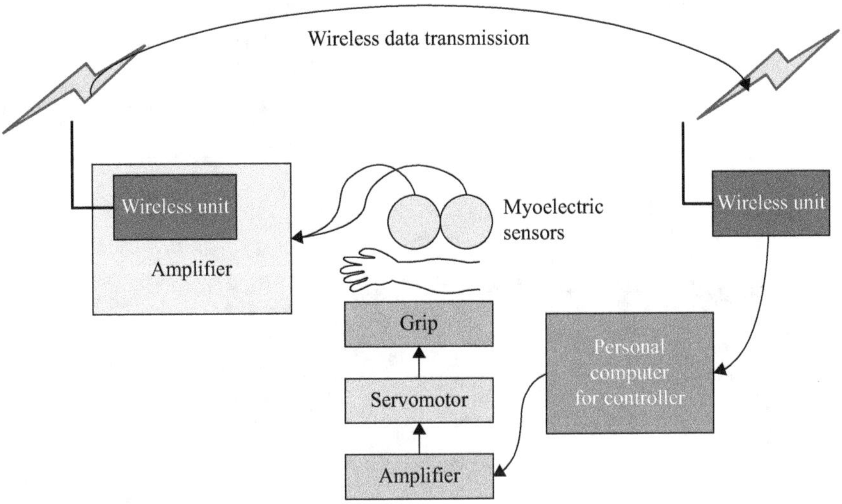

Figure 8.25 Rehabilitative training robot using myoelectric sensors

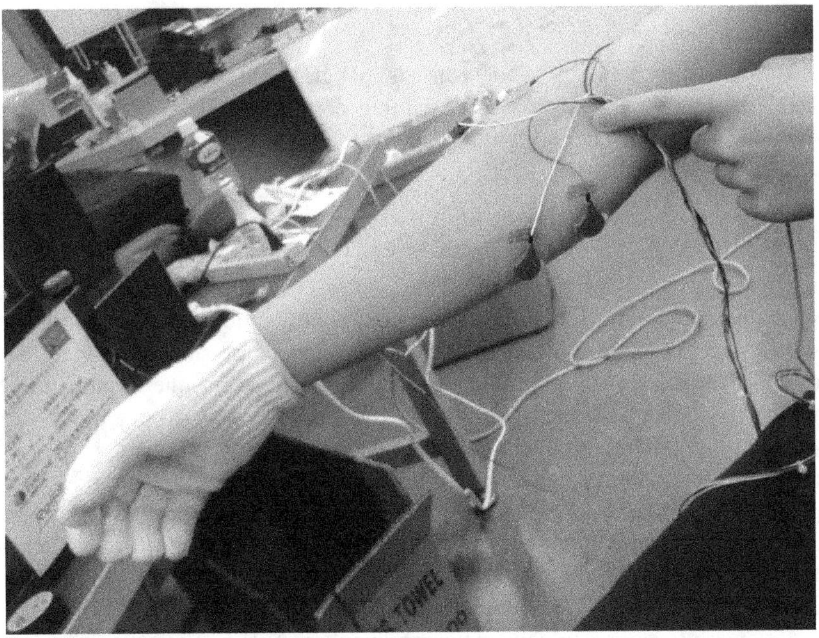

Figure 8.26 Myoelectric sensors

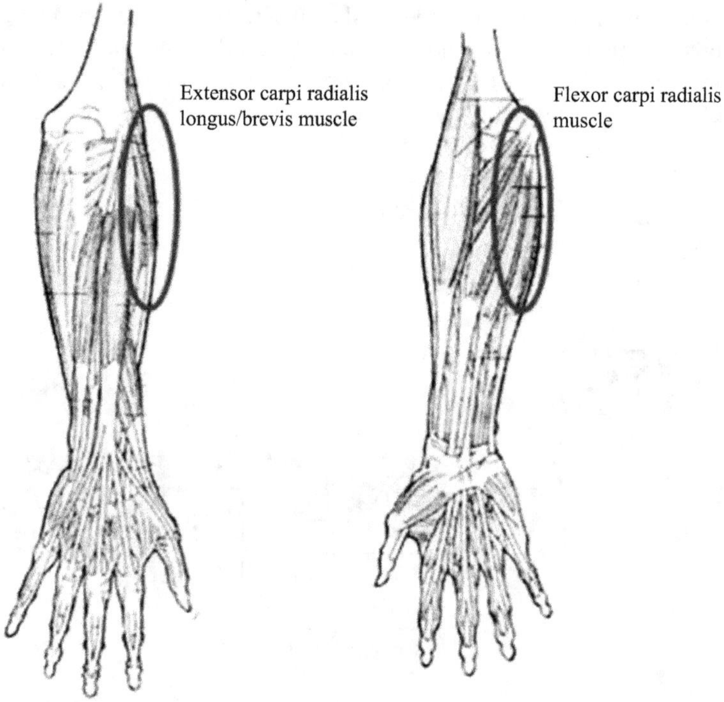

Figure 8.27 Extensor carpi radialis longus/brevis muscle and flexor carpi radialis muscle

The rehabilitative training robot consists of grip, actuator, biological signal detector, biological signal processor, training controller, and both-wrist system. Grip is held by the patient during training. The training movement is generated by a servomotor as the actuator.

8.2.3 System development

The system was first designed by CAD as shown in Figure 8.29. Through adjustments of manufacturing for safe use, the wrist rehabilitation robot was developed as shown in Figure 8.24. The grip and computer display are shown in Figure 8.30. The robot is small desk size and is easy to carry. Therefore, the training can be performed in a small space in the hospital and at home. The passive mode is the mode of palmar and dorsal flexion motions started by the start button, as shown in Figure 8.30. The constant torque mode is the mode in which patients operate the grip under a constant torque. The autonomous training mode has two main functions. One function is that the patient operates an arbitrary grip A of one hand in Figure 8.31. The palmar and dorsal flexion motions of the other hand's grip B and grip A itself are started autonomously. The other function is that the patient operates an arbitrary grip to palmar flexion direction A; the palmar flexion motion

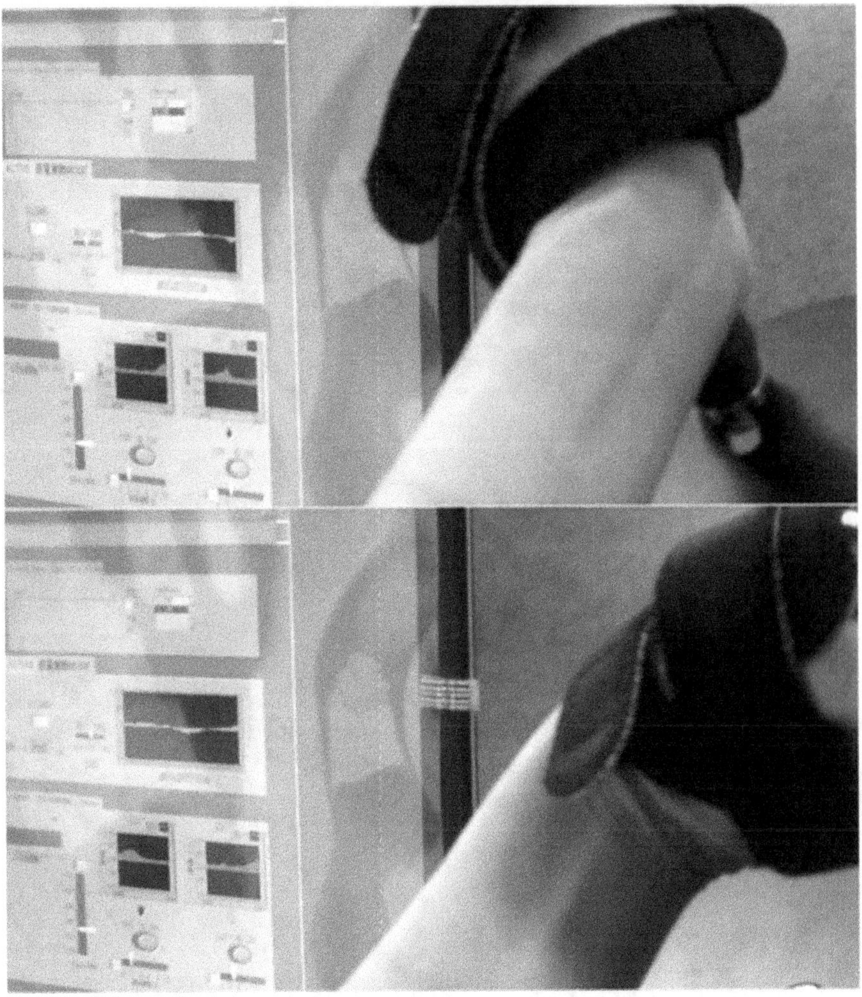

Figure 8.28 Developed rehabilitation robot in an excited condition

B is started as shown in Figure 8.32, and the dorsal flexion motion is started when the patient operates it to the dorsal flexion direction.

8.2.4 Interface system

The rehabilitation robot system has been tested in a hospital. The system has a user-friendly screen for patients, with easily recognizable touch panels; therefore, it can avoid operational error when used by patients on their own. The training data can be stored for each patient, and re-start training is possible when a patient stops training. In addition, the height of the desk can be adjusted smoothly manually according to the patient's height. A limit switch of maximum flexion angle of the patient and an emergency switch to stop the servomotors, are equipped with the

Figure 8.29 CAD design of the new rehabilitative training robot for hospital use

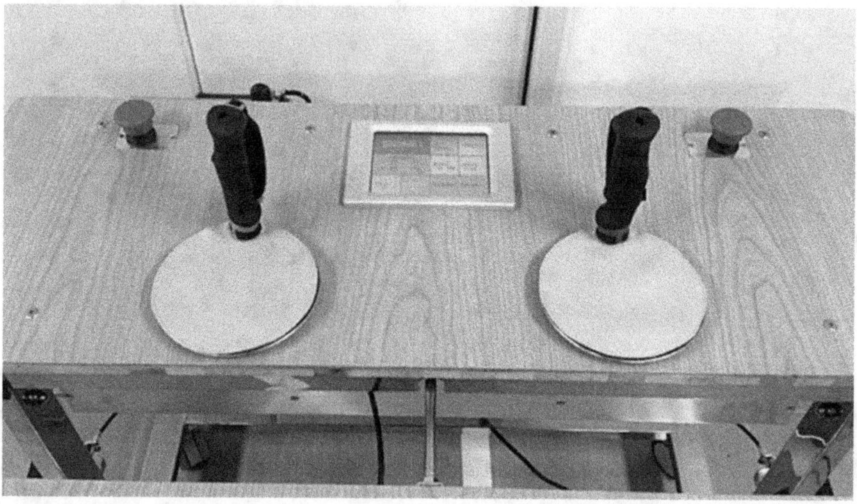

Figure 8.30 Grip and computer display

system for safety. The most effective function is 'mirror effect training'. It is found that mirror effect motion is effective in patient rehabilitation. The control system is equipped under each circular grip of the hands, and the mirror effect control, motion adjustable to the patient's state, is conducted; therefore, the paretic hand can learn how to move by the real time teaching data of the non-paretic hand.

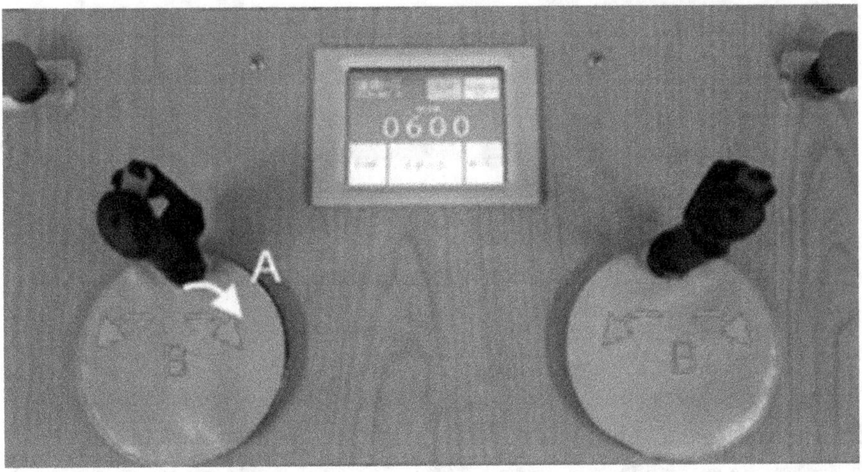

Figure 8.31 Autonomous training mode 1

Figure 8.32 Autonomous training mode 2

The same time interval of flexion training of both wrists at different rotation speeds can be realized by the control system.

8.2.5 System test

This system was tested in a hospital and feedback from the experimental trial was obtained. The screen displayed by the feedback is shown in Figure 8.33. Reducing the count number in the screen enables patients to maintain the motivation of rehabilitation.

However, because of the monotonous movements of the robot, patients can fall asleep in the middle of training, and reduction of motivation was seen. Thus, an

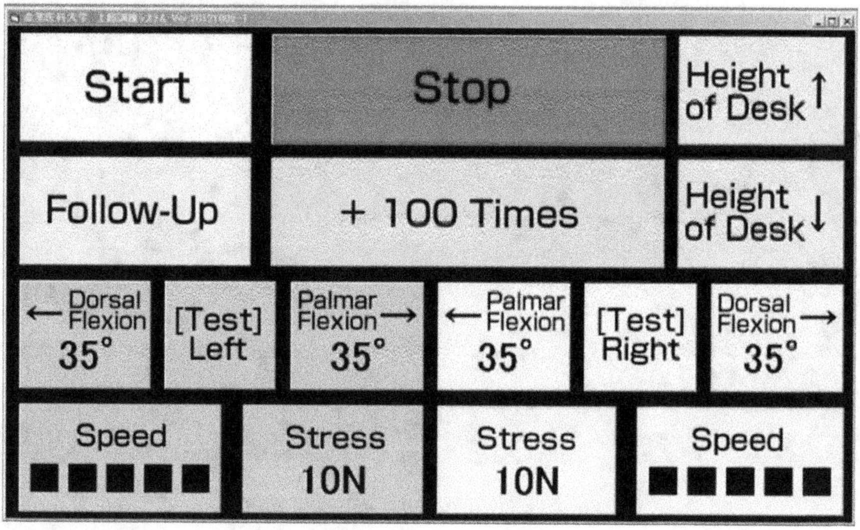

Figure 8.33 Operation screen

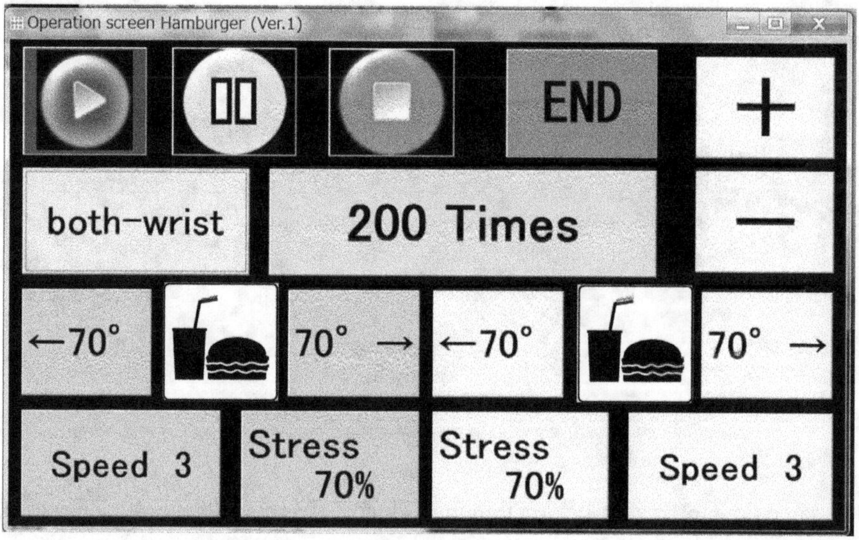

Figure 8.34 Improved operation screen

incitement is needed to avoid sleeping during training. Therefore, the author and his colleagues developed a visual interface as a new system that can enhance the training motivation. The improved operation screen is shown in Figure 8.34. As the patient gradually performs the number of repetitions in training, the picture changes, and is visually stimulating to maintain the patient's training motivation. In addition,

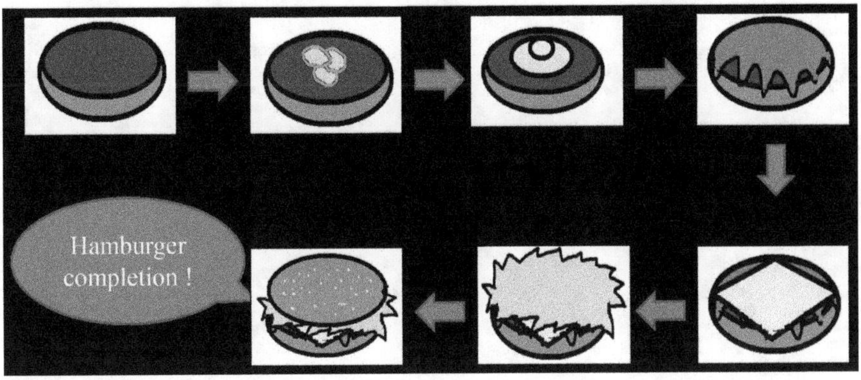

Figure 8.35 Image construction

periodic sound prevents a patient's sleep during training. The image construction is shown in Figure 8.35. The patient makes a hamburger. In addition, the author and his colleagues have increased the variations of the image construction.

8.2.6 Conclusion

The practical wrist rehabilitation robot has been developed and is used in hospitals. The robots are introduced and applied to several hospitals, and the author and his colleagues obtain clinical data to complete the research. In addition, the author and his colleagues improve more functions according to the requirements from medical doctors, therapists, and patients.

A part of study was supported by Grant-in-Aid for Japanese Scientific Research (B) and FAIS research.

Chapter 9
Advances in sustainable energy systems

Development of ocean energy power generation has long been considered potentially the next generation of renewable energy resources.

The author has begun research on the construction of cluster-based offshore ocean energy generation technology, as shown in Figure 9.1. Specifically, this includes the development of a hybrid of wind power, wave power, and tidal power generation, and the development of marine equipment using resilient vibration blades (this type of power generation would be ideally suited for applications such as local fisheries), and the development and construction of power plant maintenance systems. All efforts are being made to promote this concept, and the author and his colleagues are eager to proceed with a prototype system offshore near Nagasaki, Japan.

For example, the wave power generator shown in Figure 9.2 can convert waves into dynamic rotation and produce electrical power. Figure 9.3 shows the mechanism for tidal current power, which converts varying water current forces into lateral forces; this is a reverse mechanism of the robotic fish's oscillating fin actuator driving mechanism.

Related to the offshore sustainable energy system, the author and his colleagues also plan to develop an AUV and ROV ocean-based test centre as shown in Figure 9.4. The author plans to make prototypes of such technologies in order to be able to contribute to the standardization of the mechanisms in regard to the robotics and mechatronics required.

The author and his colleagues succeeded the real-time monitoring of offshore floating wind generator in Nagasaki, Japan, by using ROVs shown in Figures 9.5–9.7.

Finally, the author expects that the above technological and human resource development will provide a base or prototype on which an ocean energy industry could potentially be built. Also the author is developing the IoT systems of production and maintenance for the ocean energy industry.

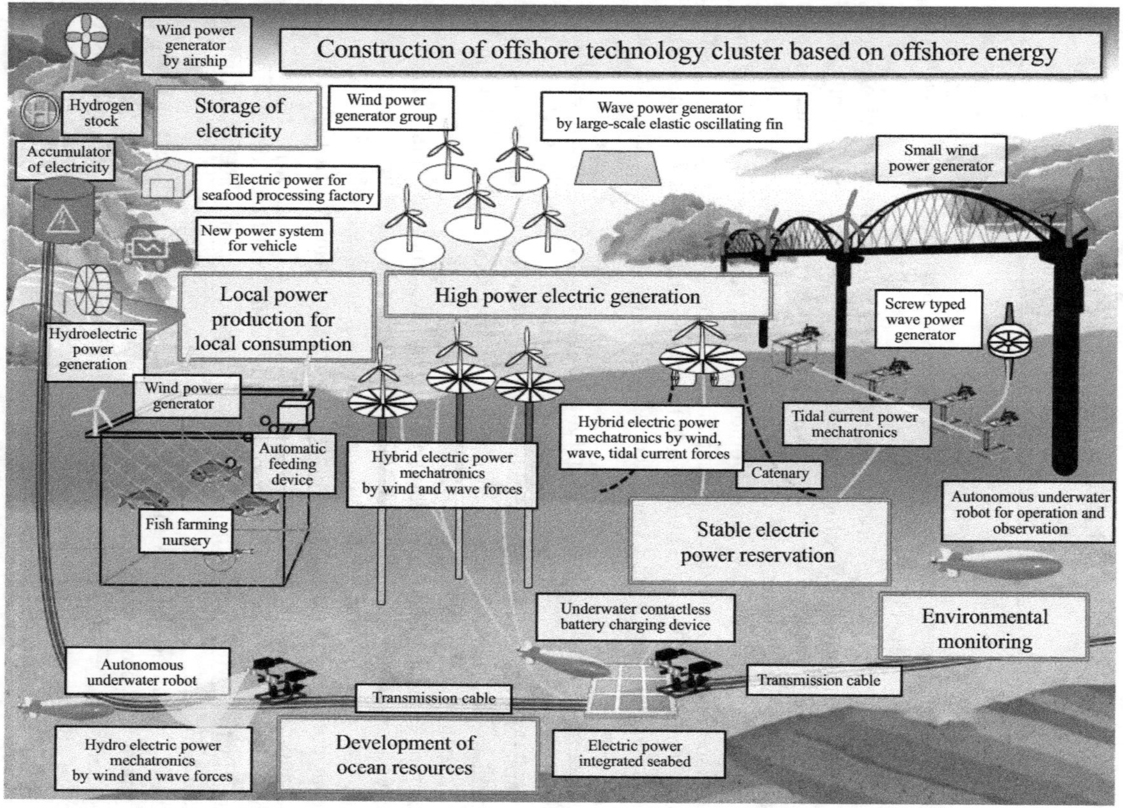

Figure 9.1 Design of an offshore energy production system in Japan

Figure 9.2 Wave power generator

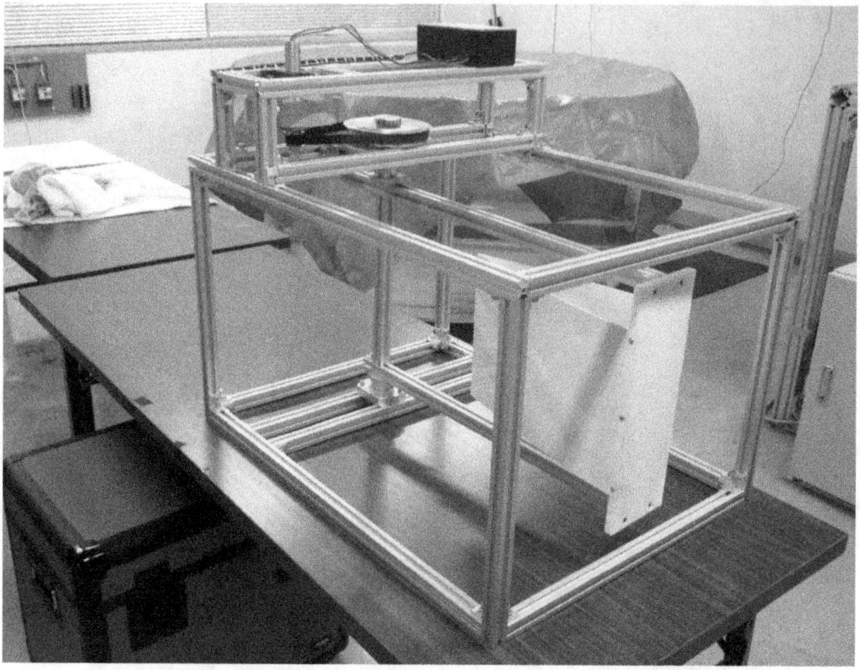

Figure 9.3 Tidal current power mechanism

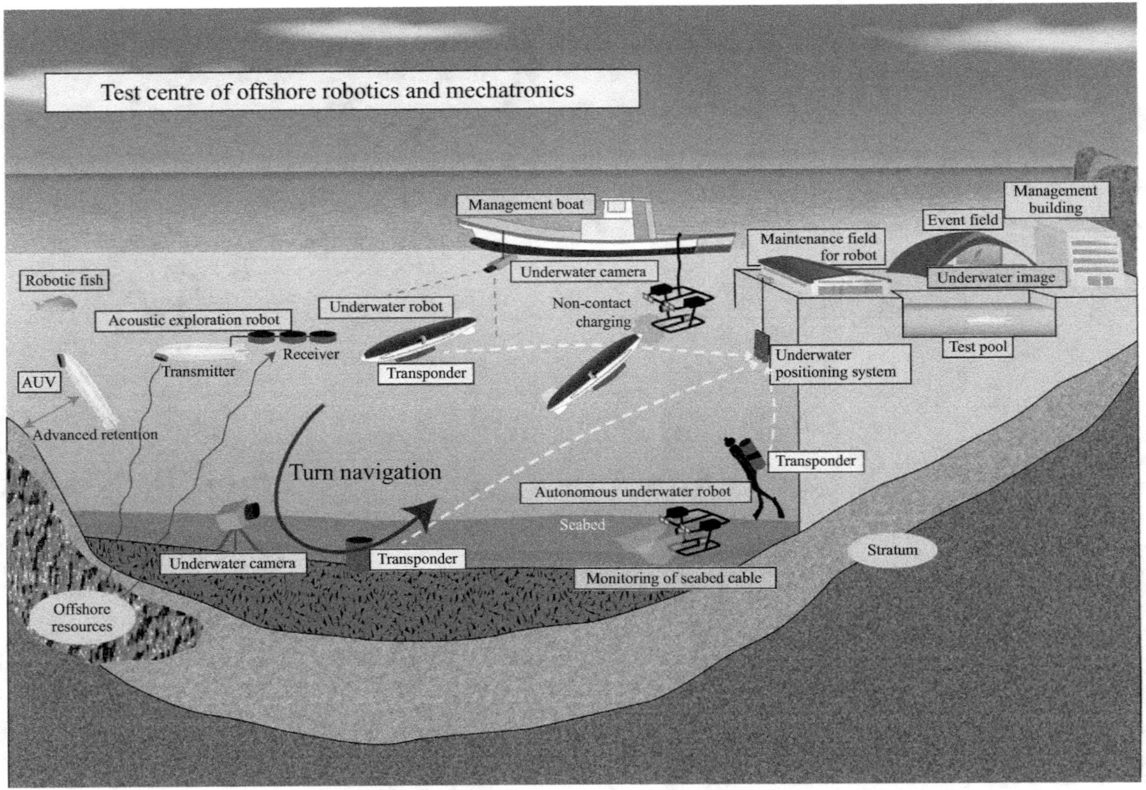

Figure 9.4 Offshore test concept for underwater robotics and mechatronics

Figure 9.5 Offshore floating wind generator in Goto, Nagasaki, Japan

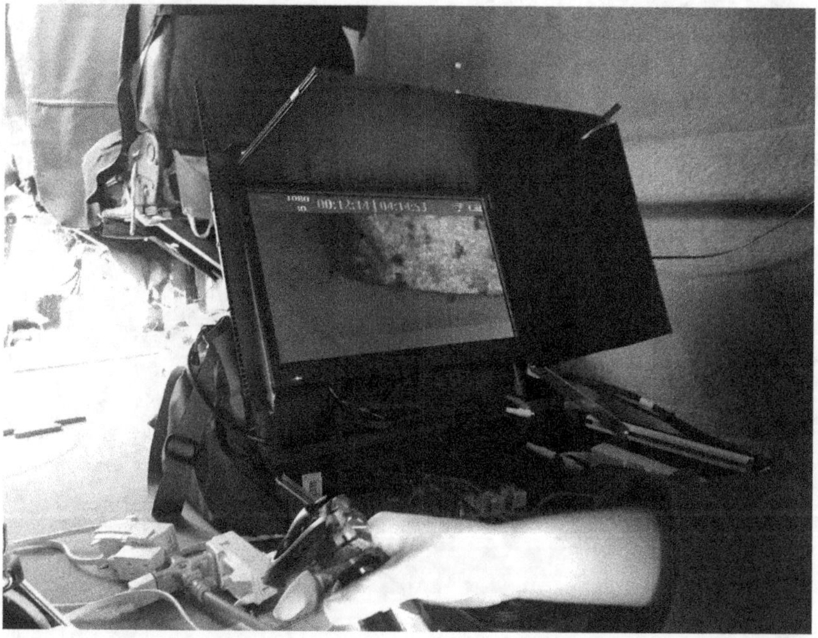

Figure 9.6 Real-time monitoring at the bottom of floating cone of offshore wind generator

Figure 9.7 ROVs and members of experiments

Further Reading

Ikuo Yamamoto, Propulsion System with Flexible/Rigid Oscillating Fin, *IEEE Journal of Oceanic Engineering*, Vol. 20, No. 1, 1995.

Ikuo Yamamoto, US Patent: SUBMERSIBLE VEHICLE HAVING SWINGING WINGS, 18.7.2000 Patent No. 6089178, 2000.

Ikuo Yamamoto, Research on Flexible Oscillating Fin Propulsion System and Robotic Fish, *Proc. IFAC CAMS*, 2001.

Ikuo Yamamoto, Robust and Non-linear Control of Marine System, *International Journal of Robust and Nonlinear Control*, Vol. 11, No. 13, 2001.

Ikuo Yamamoto, EU Patent: FISH-SHAPED UNDERWATER NAVIGATION BODY, CONTROL SYSTEM THEREOF, AND AQUARIUM 30.04.2002 Patent No. JP0204306, 2002.

Ikuo Yamamoto, Research and Development of Autonomous Underwater Vehicles, *Proc.7th UUVS*, UK, No. 16, 2005.

Ikuo Yamamoto, Research on Bio-Maneuvering Type Underwater Vehicle, *Proc. UUST05 USA*, No. 3, 2005.

Ikuo Yamamoto and Yuuzi Terada, Automatic Control and Guidance Method of ROV with Umbilical Cable, *Proc. IFAC CAMS*, 2007.

Aryeh Marks, James F. Whidborne, Ikuo Yamamoto, Control Allocation for Fault Tolerant Control of a VTOL Octorotor, *Proc. UKACC*, UK, 2012.

Ikuo Yamamoto, Autonomous Underwater Vehicles, Compendium of Ship Hydrodynamics, Chapter16, ENSTA Paris Tech, 2015.

Ikuo Yamamoto, Research on Next Autonomous Underwater Vehicle for Longer Distance Cruising, *Proc. IFAC NGCUV*, 2015.

Swimming robotic fish can be found in: https://drive.google.com/open?id=0B-n6yBhMARKebjgtd0tlazctczA

Index

www.ingramcontent.com/pod-product-compliance
Lightning Source LLC
LaVergne TN
LVHW011804200926
848484LV00003B/780